China and India, 2025

A Comparative Assessment

Charles Wolf, Jr., Siddhartha Dalal, Julie DaVanzo,
Eric V. Larson, Alisher Akhmedjonov, Harun Dogo,
Meilinda Huang, Silvia Montoya

Prepared for the Office of the Secretary of Defense

Approved for public release; distribution unlimited

 RAND NATIONAL DEFENSE RESEARCH INSTITUTE

The research described in this report was prepared for the Office of the Secretary of Defense (OSD). The research was conducted in the RAND National Defense Research Institute, a federally funded research and development center sponsored by OSD, the Joint Staff, the Unified Combatant Commands, the Navy, the Marine Corps, the defense agencies, and the defense Intelligence Community under Contract W74V8H-06-C-0002.

Library of Congress Cataloging-in-Publication Data

China and India, 2025 : a comparative assessment / Charles Wolf ... [et al.].
    p. cm.
  Includes bibliographical references.
  ISBN 978-0-8330-5042-7 (pbk. : alk. paper)
  1. Economic development—China—Forecasting. 2. Economic development—India—Forecasting. 3. China—Population. 4. India—Population. 5. Technological innovations—China. 6. Technological innovations—India. 7. China—Armed Forces—Appropriations and expenditures. 8. India—Armed Forces—Appropriations and expenditures. I. Wolf, Charles, 1924- II. Rand Corporation.
  HD77.5.C6C45 2011
  338.951054—dc23
                                                                    2011029057

Published 2011 by the RAND Corporation
1776 Main Street, P.O. Box 2138, Santa Monica, CA 90407-2138
1200 South Hayes Street, Arlington, VA 22202-5050
4570 Fifth Avenue, Suite 600, Pittsburgh, PA 15213-2665
RAND URL: http://www.rand.org/
To order RAND documents or to obtain additional information, contact
Distribution Services: Telephone: (310) 451-7002;
Fax: (310) 451-6915; Email: order@rand.org

Preface

China and India will exercise increasing influence in international affairs in the coming decades. As prominent members of the G-20, their influence will be manifest in the global economy, in global politics, and in the global security environment. Each country's role on the world stage will also be affected by the progress that it makes and by the competition and cooperation that develop between them.

The research described in this monograph focuses on the progress China and India seem likely to achieve from 2010 through 2025, as well as on some of the major problems they may encounter along the way. This research consists of a comparative assessment of their prospects in this period in four domains: demography, macroeconomics, science and technology, and defense spending and procurement. In each domain, the assessment seeks answers to these questions: Who is ahead? By how much? and Why? Often the answers are quantitative, sometimes they are more qualitative, and sometimes they are inconclusive. The monograph concludes with implications for policy and for further research.

In view of this scope, this document should be of interest to decisionmakers and analysts in the executive branch, in Congress, and among the larger public.

This research was sponsored by the Director of Net Assessment in the Office of the Secretary of Defense and conducted within the International Security and Defense Policy Center of the RAND National Defense Research Institute, a federally funded research and development center sponsored by the Office of the Secretary of Defense, the Joint Staff, the Unified Combatant Commands, the Navy, the Marine Corps, the defense agencies, and the defense Intelligence Community.

For more information on the RAND International Security and Defense Policy Center, see http://www.rand.org/nsrd/about/isdp.html or contact the director (contact information is provided on the web page).

Contents

Figures

Tables

Summary

In the past century, China and India have experienced frequently rivalrous relations, including two occasions of military conflict in 1956 and 1982, sharp changes in the issues and venues of their rivalry, and sometimes quite different stances toward the United States and its policies. These circumstances provide a backdrop for our report, although our approach is more narrowly focused, while also looking forward to the two countries' future prospects rather than to their histories.

The purpose of this document is to assess the prospects of India and China through 2025 in four domains: demography, macroeconomics, science and technology, and defense spending and procurement. We seek to answer these questions: Who's ahead? By how much? and Why? As the second question implies, we strive for quantitative answers as much as possible. In the process, we try to assess the balance between advantages and disadvantages that China and India will possess 15 years hence. This balance is relevant for potential cooperation between the two countries, no less than for their potential competition and rivalry. Although our focus is on quantitative answers, we repeatedly acknowledge the uncertainties created for the assessment by such qualitative unknowns as whether or not each country may encounter internal civil unrest, political disruption, external conflict, or natural disasters.

Demography

Chapter Two begins the comparative assessment by examining the demographic balance. China and India are the world's two most populous countries. India's current rate of population growth is about twice that of China (1.55 percent annually, versus 0.66 percent for China), and its total population will equal China's in 2025 (about 1.4 billion in each country), thereafter exceeding China's. The Indian population will continue increasing through at least 2050, while China's will peak at about 1.5 billion in 2032, declining thereafter.

From the standpoint of economic competition between the two countries, the age composition of their populations is more significant than their aggregate size. India's prime-working-age population will overtake that of China in 2028. Moreover, reflecting the changing age composition of their populations, the two countries will experience very different patterns in their overall dependency ratios—that is, the ratio of the young and the elderly to those of prime working age. (The dependency ratio concept assumes that, on average, people aged 15–64 produce more than they consume, while the opposite is true for those who are younger and older. Rising dependency ratios are generally viewed as an impediment to economic performance, while falling ratios are considered an advantage.) Although India's overall dependency ratio is currently higher than China's, the ratio will be rising rapidly in China in the next two decades, while it will be declining in India.

Numerous other factors will affect the balance of demographic advantages and disadvantages, including the health, education, gender composition, and migration propensities of the respective populations. For example, China's population is generally healthier than India's, and China has the benefit of a more developed health care system. On the other hand, China's population is aging more rapidly than India's, in the sense that the elderly are becoming an increasingly larger proportion of China's population. India will have a lesser cost burden from this source because of its younger population.

China's population also has higher average levels of literacy and education than India's. If India can successfully meet this challenge by investing in human capital, it may be able to turn a disadvantage into an advantage through productive employment of its growing pool of younger workers.

The bottom-line answer to the "Who is ahead?" question as it relates to demography is evidently in India's favor. However, whether India's several demographic advantages—increasing numbers, younger age cohorts, declining dependency ratios—will be a dividend or drag on future economic growth will depend on the extent to which productive employment opportunities emerge from an open, competitive, innovative, and entrepreneurial Indian economy. Conversely, whether China's several demographic disadvantages—rapidly aging population, rising dependency ratios, rising health costs for the elderly, sharp gender imbalances—will be a drag or a dividend will depend on the extent to which these demographic circumstances provide a stimulus to improving technology and to raising the skill and productivity of a shrinking labor force.

Macroeconomics

In Chapter Three, we assess the macroeconomic balance between India and China through a meta-analysis of 27 recent studies of the two countries' prior economic growth and their forecasted growth through the 2025 period. The studies were selected from a larger set of 47 studies screened on the basis of the scope and reliability of the data needed for the meta-analysis. The studies, published between 2000 and 2008, were from three different types of institutions: academic, business, and international organizations. The pooled data enabled comparisons to be made between China's and India's forecasted economic performance through 2025 in terms of four salient indicators: growth of capital, growth of employed labor, growth of total factor productivity, and growth of gross domestic product (GDP).

What is striking about the results is the narrow margins between the paired China-India comparisons. The forecasted average annual GDP growth rates in 2020–2025 are approximately the same: China at 5.7 percent, India at 5.6 percent. The corresponding maximum GDP growth rates of the forecasts are 9.0 percent for China and 8.4 percent for India, and their paired minimum growth rates are 3.8 percent and 2.8 percent, respectively. Estimates of the other three growth indicators (capital, labor, total factor productivity) show slightly larger differences.

The meta-analysis also included comparisons among the three separate clusters, covering 11 academic, 9 business, and 7 international organizations. The business cluster's forecasts are distinctly more optimistic about India's growth prospects and relatively pessimistic about China's, forecasting an average Indian growth rate of 6.3 percent and an average Chinese growth rate of 4.7 percent for the 2020–2025 period. The two other clusters (academic and international organizations) reverse this order, with markedly higher growth estimates for China than for India. We conjecture that an expectation of a more favorable business environment in India—for example, relating to the rule of law and protection of property rights—might account for this difference.

To reflect as well as to bound the uncertainties embedded in the meta-analysis forecasts, our assessment shows the GDP comparisons between India and China that result from five differing paired scenarios of their respective high, low, and average growth rates. Only in the scenario that posits the high-growth parameter for India and low-growth for China does India's GDP in 2025 approach that of China. In this scenario, India's GDP in 2025 is $12.3 trillion and China's is $13.8 trillion, employing purchasing power parity (PPP) rates to convert rupees and renminbi, respectively, to constant U.S. dollars.

We conclude that, concerning forecasted economic growth, our assessment places the two countries at equivalent rates, but with China's aggregate GDP likely remaining substantially larger than India's through 2025, as is currently their comparative status.

Science and Technology

We assess the science and technology (S&T) balance between India and China in Chapter Four. The assessment focuses on several indicators of S&T inputs and two output indicators. The input indicators include both financial and human resources. The financial input indicators involve spending on research and development (R&D). We focus on gross expenditures on R&D (GERD) as a percentage of GDP, as well as GERD's four components: higher education R&D spending (HERD), business R&D spending (BERD), government R&D (GOVERD), and private, nonprofit organizations' R&D spending (PNPERD). The human resource input indicators are the number of doctoral degrees in engineering, life sciences, physical sciences, computer science, mathematics, and agriculture.

As output indicators, the assessment compares (1) publications in refereed scientific journals and (2) patents (especially triadic patents) produced by authors and inventors from China and from India. We acknowledge that these indicators are incomplete: For example, innovations and improvements in production and management practices often occur that are not reflected in either scientific publications or patents. Despite their limitations, these indicators are used in the assessment of Chapter Four to compare India's and China's recent S&T accomplishments and to develop a simulation model for projecting their future trajectories.

China currently has the world's third-largest GERD (after the United States and Japan). Also, the business component (BERD), which may have the greatest early effects on productivity, has increased from 0.25 percent of China's GDP in 1996 to more than 1 percent in 2006. China's GERD has subsequently risen further in absolute amounts and as a share of China's GDP. India's GERD is 0.8 percent of GDP; it is expected to triple in the next five years and to continue to rise through the 2025 period.

China currently graduates 70 percent more engineers annually than does India (600,000 and 350,000, respectively). However, there are questions about the reliability and comparability of these aggregate figures, and another difficulty arises in assessing the quality of similarly credentialed engineers in the two countries. As an example, according to a survey of multinational businesses, the quality ("employability") of graduate engineers from China is 60 percent less than that of graduate engineers from India.

The simulation model described in Chapter Four uses the input variables mentioned above, along with cost and output parameters. The parameters are sometimes based on current levels prevailing in India and China and sometimes based on current levels in South Korea, on the assumption that the parameter values in China and India will converge over the next 15 years to the higher levels prevailing in South Korea in 2008. Our assessment includes several simulation scenarios with differing combinations of these parameter values and differing degrees of optimism and pessimism about prospective S&T developments in India and China.

Whether outputs are registered in terms of full-time science and engineering researchers, holders of doctoral S&E degrees, triadic patents, or journal publications, the forecasted answer to the "Who is ahead?" questions is that our estimates for China exceed those for India by wide margins. The simulation estimates of China's researchers and S&E journal publications in 2025 exceed those of India by factors of 8 and 13, respectively. Only in the scenarios in which we adopt the qualitative discount cited above for China's graduate engineers and their imputed productivity do these factors diminish substantially, falling to 1.5 and 1.7, respectively.

Spending on Defense and Defense Procurement

Comparing spending on defense and defense procurement in India and China involves problems of data reliability and comparability that are no less difficult than those encountered in the preceding S&T comparisons. The assessment in Chapter Five addresses these problems, as well as the additional problem presented by identifiable gaps in the defense spending and procurement data for both countries. While these gaps are evident in both cases, they are distinctly larger in China. The approach adopted in Chapter Five builds on each country's official data to arrive at estimates of their total expenditures on defense and on defense procurement, and to express these as shares of their respective GDPs.

Forecasts of these expenditures through 2025 are made using two methods. One method is based on a continuation of recent year-over-year real growth rates of defense spending, while the second method assumes that defense spending is a fixed share of GDP, thereby linking the defense spending estimates to the GDP forecasts provided in Chapter Three. The first method yields substantially higher forecasts for defense spending than the second, resulting in budget and GDP shares so high that they would likely be politically unacceptable in both countries. Each of the two methods was used to generate three different estimates for China and India, representing optimistic ("high"), pessimistic ("low"), and moderate ("best") scenario assumptions.

According to the first method, our "best" estimate for India's defense spending in 2025 is between $94 billion and $277 billion in 2025, in constant dollars depending, respectively, on whether market exchange rates or PPP conversion rates are used. The corresponding "best" estimates for China are between and four and seven times those for India. As noted earlier, the forecasts resulting from the second, GDP-based method are appreciably lower, lying between $82 billion and $242 billion for India, and between two and three times these amounts for China.

Turning to spending for defense procurement, we employ a single estimation method analogous to the year-over-year method cited above for estimating defense spending through 2025. The method posits a high and fixed (12.8 percent) annual growth of procurement spending for both countries from 2009 levels. Our resulting estimates for India's defense procurement in 2025 are between $63 billion and $186 billion (in constant dollars), depending on whether market exchange or PPP conversion rates are used to convert rupees to dollars. The corresponding "best" estimates for China are about 2.6 times and four times these amounts.

Observations and Implications

Chapter Six concludes with observations about the four dimensions of the assessment and with implications that may be drawn from the assessment. We reiterate the abundant sources of uncertainty surrounding our quantitative estimates and advise caution in treating our forecasts of economic growth, scientific and technological development, and defense spending as other than suggesting boundary conditions. While recognizing the uncertainties, our answers to the original questions about "Who is ahead?" and about the respective advantages and disadvantages of India and China can be briefly summarized:

- The demographic assessment suggests several distinct advantages for India (these are delineated in Chapter Two).
- The macroeconomic assessment suggests that the economic growth competition between India and China may be considerably closer than might otherwise be expected.
- In S&T, China's margins over India are likely to be substantial, deriving largely from the currently prevailing disparities between them that, in absolute terms, are likely to grow.
- In defense spending and procurement, a similar pattern is likely to emerge: The two countries show wide disparities in their current spending levels and, in absolute terms, these are likely to grow substantially over the next 15 years.

An important implication follows from the multiple high-versus-low/optimistic-versus-pessimistic scenarios described in our assessment. India and China, by adopting or failing to adopt suitable policies, can affect significantly the probabilities that one or another of the alternative scenarios materializes, thereby altering the balance of advantages and disadvantages between the two countries. For example, if India follows effective economic and social policies, its favorable demographic trends will result in a significant "dividend" for the economy's growth; conversely, if China's policies were to fall short of compensating for the adverse demographic trends it faces, the result will be a significant "drag" on its economic growth.

Also, though perhaps to a more limited extent, policies pursued by the United States and other "third" parties may be able to affect this balance. From this standpoint, an actionable inference can be drawn: Identify which among the multiple scenarios sketched in our assessment seems preferable, and develop a portfolio of policies conducive to enhancing the probability of that (or those) scenario(s) emerging over the next 15 years.

Explicating the specific policies and their effects in altering our forecasted outcomes is worthy of further attention, as well as beyond the purview of this study. Nonetheless, we suggest the following proposition: Prospects for India to pursue policies that will enhance its competitive position vis-à-vis China are better than are the reverse prospects. This is because India's political-economic system entails at least a moderately greater degree of economic freedom than does China's, and this provides an environment more conducive to entrepreneurial, innovative, and inventive activity that may favor India's position in the long-term competition between the two countries.

Acknowledgments

We are indebted to our two formal reviewers: RAND colleague Michael Mattock and Enders Wimbush of the Hudson Institute. Their detailed and constructive comments were invaluable in helping us make numerous revisions and improvements to the document. We are also indebted to David Epstein, deputy director of the Office of Net Assessment in the Department of Defense, for his many comments and suggestions. We have benefited as well from careful and thorough editing by James Torr.

These acknowledgments are, of course, made with the usual absolutions to all of those mentioned above for what appears in the final text.

Abbreviations

ANU	Australian National University
APEC	Asia-Pacific Economic Cooperation
BERD	business expenditures in R&D
CBR	crude birth rate
CDR	crude death rate
DG	Demand for Grants
DIG	Defense Industry Group
EPO	European Patent Office
EU	European Union
FTR	full-time researcher
GDP	gross domestic product
GERD	gross expenditures in R&D
GOVERD	government expenditures in R&D
HERD	higher education expenditures in R&D
HRST	human resources for science and technology
IDB	U.S. Census Bureau International Data Base
INR	Indian rupees
JPO	Japan Patent Office
LCU	local currency unit
LEB	life expectancy at birth

MXR	market exchange rates
OECD	Organisation for Economic Co-operation and Development
PAP	People's Armed Police
PLA	People's Liberation Army
PNPERD	private nonprofit expenditures in R&D
PPP	purchasing power parity
R&D	research and development
RMB	Chinese renminbi
S&E	science and engineering
S&T	science and technology
TFP	total factor productivity
TFR	total fertility rate
UNESCO	United Nations Educational, Scientific, and Cultural Organization
USPTO	U.S. Patent and Trademark Office
WHO	World Health Organization

CHAPTER ONE

Objectives, Background, Context

The purpose of this document is to assess the relative levels, attainments, and prospects of China and India through 2025 in four domains: demography, macroeconomics, science and technology, and defense spending and procurement. We also seek to identify impediments and constraints that each country will confront in these domains through the next 15 years. Simply stated, we try to answer, or at least shed light on, the following questions: Who is ahead? By how much? and Why?

As the second of these questions implies, we mainly, although not exclusively, strive for quantitative answers. At the same time, we repeatedly acknowledge the many crucial qualitative factors that may decisively influence the quantitative answers we seek, such as the occurrence of civil unrest, political disruption, external conflicts, or natural disasters. But while these omitted factors may have important effects on the four dimensions we focus on in this study, these effects will not necessarily change the positions of China and India relative to one another. For example, China may be neither more nor less subject to civil unrest or political disruption than India.

The objectives of this study bring to mind a magisterial study completed a decade ago, *Protracted Contest: Sino-Indian Rivalry in the Twentieth Century* (Garver, 2001), that surveys the complex relations and interactions—including two occasions of military conflict in 1956 and 1962—between China and India in the past century. John Garver's book focuses exhaustively on the history, culture, politics, diplomacy, and geography of the two countries' rivalrous relations across a wide range of issues, including especially those relating to Tibet, Pakistan, Burma, Nepal, other developing nations, and the Indian Ocean.

Garver's focus is both relevant to and different from that of our study. Its relevance lies in the fact that the domains addressed in our assessment define and quantify some of the major capabilities as well as the constraints—both the advantages and disadvantages—that the two countries would bring to the continued rivalry and the long-term competition between them in the 21st century, through 2025.

However, our focus also differs from Garver's because the capabilities and constraints that we address may affect potential cooperation between India and China, as well as their rivalry and competition. Further, we explore how the capabilities and constraints may also influence the United States and other countries in formulating policies toward both countries.

India and China command special attention, not just in Asia (for example, in the Asia-Pacific Economic Cooperation [APEC] forum and the Council for Security Cooperation in the Asia Pacific [CSCAP]) but also in the global economy and in the biennial, multicountry consultative summit meeting known as the G-20, which is acquiring growing importance in the international arena. China and India are the world's two most populous countries. They have sustained the world's highest annual gross domestic product (GDP) growth rates over the past decade—9 percent for China and 6 or 7 percent for India. The two countries have been among the world's most successful in weathering the challenges of the global economy's Great Recession since 2008. China has accomplished this through a combination of a large government stimulus program (as a share of its GDP twice as large as that of the United States) and an effective infrastructure-building program. India's similarly successful efforts in sustaining rapid growth despite the global recession have been due to its lesser dependence on exports to drive its economy and an expansion of domestic demand.

The two countries arguably have the greatest influence and leverage among the ten emerging-market countries in the G-20. Their joint influence has been decisive in aborting the World Trade Organization's Doha Development Round of negotiations on trade liberalization, as well in the failure—whether for good or ill—of the 2009 UN Climate Change Conference (also known as the Copenhagen Summit). China has become the world's largest source of net capital outflows (Wolf et al., forthcoming). India's popularity as a destination for foreign capital inflows is rapidly increasing, and India is the world's largest recipient of foreign outsourcing of computer-based services.[1] China and India are both heavily dependent on imported oil: They are the world's second- and fourth-largest importers, respectively. Shifting to a very different realm, China is the most aggressive opponent of the Dalai Lama and Tibetan autonomy, while India is their most vigorous supporter.

The prominence of India and China in all of these issues is indisputable. But the relevance of the four domains to each of the above issues varies. The two countries' **demographics** have some bearing on most of these issues. China's and India's likely **economic growth** trajectories affect and are affected by most of them. **Science and technology** in the two countries will affect their respective competitive positions in several of these issues, as will their respective performance in the domain of **defense spending and procurement**.

While the four domains of our assessment are thus pervasively important and timely, they are not exhaustive or dispositive. Our assessment ignores, or at best considers tangentially, numerous circumstantial and institutional factors that may often dominate the four domains addressed in affecting many issues, whether of rivalry or of

---

[1]   FactSet Mergerstat, 2009; Wikipedia, undated.

cooperation, between India and China. These exogenous factors include, for example, the evolution of each country's domestic politics; the progress of democracy, pluralism, and the rule of law; civil strife; foreign military hostilities; and the policies of the United States and other third countries toward India and China. Implicitly, we assume that these other factors are unchanged, or that such changes as occur will affect India and China to the same extent. So, caution is warranted in reading predictive validity into the assessment that follows.

The analysis covered in this report has five parts.[2] Chapter Two reviews population trends in China and India. It considers whether the fact that China's population is aging much more rapidly than India's may be a "drag" on China's economic prospects and a "dividend" for those of India, or, less likely, that circumstances may arise under which the reverse effects might ensue. The chapter explores many issues on which this dividend-versus-drag question depends: the ratios of the young and the elderly to those of priming working age (dependency ratios) and of males to females (sex ratios) in the evolving age cohorts, educational demands, health conditions, the role of women, and the social implications of son preferences and gender imbalances in the societies and polities of the two countries. The chapter summarizes the balance of comparative advantages and disadvantages that demographic changes will generate for India and China in 2025.

Chapter Three is a macroeconomic assessment of India's and China's economic growth to date and their prospects through 2025. The core of Chapter Three is a meta-analysis of 27 separate independent studies, published between 2000 and 2008, of the major components of and contributors to each country's growth over the past decade: namely, growth of capital (i.e., plants and equipment), employed labor, total factor productivity (i.e., productivity of a weighted combination of capital and labor), and real growth of each country's GDP. The collective means, minima, maxima, and variances of these factors are calculated, and the 27 studies are grouped into three clusters: those done by academic authors and institutions, by business organizations (e.g., Goldman Sachs, PricewaterhouseCoopers, McKinsey), and by international organizations (e.g., the International Monetary Fund, the World Bank, the Asian Development Bank). The meta-analysis highlights several interesting contrasts, as well as some similarities, among the three clusters. Chapter Three concludes with five differing scenarios among possible pairings between average, high, and low growth forecasts derived from the three clusters.

Chapter Four organizes available data for India and China on research and development (R&D) spending in the past decade and each country's national plan for increasing it in the coming decade. The data cover four sources of funding: government, business, higher education, and private nonprofit organizations. These spending

---

[2] Several of the chapters—especially Chapters Two and Four—have been abridged from more detailed treatment of these subjects. We hope to publish these studies as separate, stand-alone monographs in the near future.

data are complemented by examining both quantitative and qualitative information on the production and mobilization by China and India of human resources: specifically, science and engineering graduates, their employment in research, their productivity (e.g., in patent awards and recorded publications), and their costs. The data are used to make projections through 2025 based on a constant-returns-to-scale simulation model driven by each country's GDP growth estimates derived from the meta-analysis summarized in Chapter Three. The model is also used to forecast each country's production of intellectual property, as represented by triadic patents and publications in refereed journals. Chapter Four concludes with projections of alternative optimistic and pessimistic scenarios relating to forecasted productivity and costs of technical researchers in India and China.

Chapter Five addresses the assessment's fourth domain—spending on defense and defense procurement. The analysis is mainly based on official government data and sources for both countries, supplemented by several other invaluable sources. The nonofficial sources help to identify gaps in official data (especially the relatively larger gaps in coverage by the Chinese sources) and to highlight differences in how the two countries define what is included in or excluded from defense spending. For example, China includes spending for police paramilitary forces in its defense spending, but India does not. The baseline 2009 estimates and forecasts to 2025 for spending on defense and defense procurement in Chapter Five are linked to the estimates of GDP and of GDP growth developed in Chapter Three, and to the estimates of real rates of growth both in GDP and in defense spending and procurement in the decade between 2000 and 2009, described in Chapter Five. These estimates are calculated in local currency units (rupees for India, renminbi for China), and U.S. dollars based on market exchange rates (MXR) and on purchasing power parity (PPP) conversion rates. Reflecting the large uncertainties inevitably involved, Chapter Five develops upper, lower, and "best" (medium) estimates for both China and India covering their total defense spending and spending on military procurement, linking these bounding estimates to the high and low GDP growth estimates from Chapter Three and the real rates of growth developed in Chapter Five, as well as to a range of specified defense-to-GDP ratios.

The 2009 baseline defense spending estimates for China are between $110 billion and $201 billion, depending, respectively, on whether PPP or MXR conversion rates are used. The corresponding baseline estimates for India are $34 billion and $101 billion, respectively. The ratio between the two countries' defense spending in 2009 was between 2.0 and 3.2, in China's favor. The ratio between the Chinese and Indian defense procurement 2009 baseline figures is between 2.6 and 4.2.

Again reflecting the abundant uncertainties implied by the several scenarios, the ratio between China's and India's defense spending in 2025 might reach as high a figure as 7.3, or might remain as low as the 2.0 low side of the 2009 baseline estimate. The corresponding ratio between the two countries' defense procurement forecasts in 2025 might be as high as 9.5 or as low as 2.6.

Finally, Chapter Six draws on Chapters Two through Five to distill answers to the original questions: Who is ahead? By how much? and Why? Chapter Six summarizes the competitive advantages and disadvantages of India and China, and their respective capabilities and constraints, through 2025. The chapter also suggests several policy implications that stem from the assessment. The conclusions and policy implications presented in Chapter Six reiterate the grounds for caution regarding the assessment, both because of the large uncertainties pervading the estimates and because of the potential influences on the "protracted contest" between the two countries that are not accounted for in our assessment.

CHAPTER TWO

Population Trends in China and India: Demographic Dividend or Demographic Drag?

Although China is currently more economically advanced than India, its population is, on average, much older than India's. Might this be a "demographic drag" that limits China's economic prospects relative to those of India? Might India, whose population is both younger and growing relative to that of China, experience a "demographic dividend" from these trends? In this chapter, we review recent demographic trends in India and China and their implications. Our focus is on the years 2020–2025; to put this period in perspective, we show data for the years 2000–2035. We mostly rely on data from the U.S. Census Bureau's International Data Base (IDB) (U.S. Census Bureau, 2010), because of the availability of both historical data (since 1950) and future projections (through 2050) and because the bureau's data on China pertain only to mainland China, which is our focus.[1] In addition, the Census Bureau recently updated these data (in June 2009 for India and in December 2009 for China), making them the most up-to-date data available at the time of our analysis.[2] Occasionally, we present data from other sources on topics not covered by the IDB.

Population Growth and Its Components

China and India are the only countries in the world with populations of more than 1 billion. According to the most recent censuses of each nation, there were 1.266 billion people in China in 2000 (National Bureau of Statistics of China, 2005) and 1.029 billion in India in 2001 (Registrar General and Census Commissioner of India, 2001). According to IDB estimates, there are now 1.330 billion people in China and 1.173 billion in India, and population growth rates have been consistently higher in India than in China since the early 1970s and will remain so for years to come. India's

---

[1]   Unless otherwise noted, the IDB is the source of the information presented here. Data from other international organizations, such as the United Nations and the World Bank, and from the countries themselves show patterns similar to those in the IDB data.

[2]   For example, the IDB's December 2009 update for China is based on a new triangulation of evidence from a variety of sources, including analysis of data from China's recent census and surveys.

population is projected to grow through at least 2050[3] (when it will be 1.656 billion), surpassing China in 2025, whereas China's population is projected to reach a maximum, of 1.395 billion, in 2026 and to decrease thereafter (Figure 2.1).

Calculating the population change for a nation is done by subtracting the number of deaths from the number of births and adding the net international migration. Although in the 20th century both nations experienced relatively large migration flows because of historical events—foreign invasion and civil war in China and the partition in India—in recent years net migration from these nations has been relatively low. According to the IDB, in 2009, India, on net, lost five persons to international migration per 100,000 population, while China lost 33 per 100,000. This leads us to focus here on other, more predictable events—i.e., births and deaths—that are currently doing much more to shape the demographics of each nation.

At present, in both China and India, the number of births considerably exceeds the number of deaths. The IDB estimates that in 2010 there will be 16.19 million births in China, resulting in a crude birth rate (CBR) of 12.17 births per 1,000 population. There are estimated to be just over half as many deaths in 2010—9.17 million—resulting in a crude death rate (CDR) of 6.89 deaths per 1,000 population. The difference between births and deaths, which is called *natural increase*, is even greater in India, where an estimated 25.03 million births are expected in 2010, resulting in a

Figure 2.1
Total Population Sizes, and China and India, 2000–2035

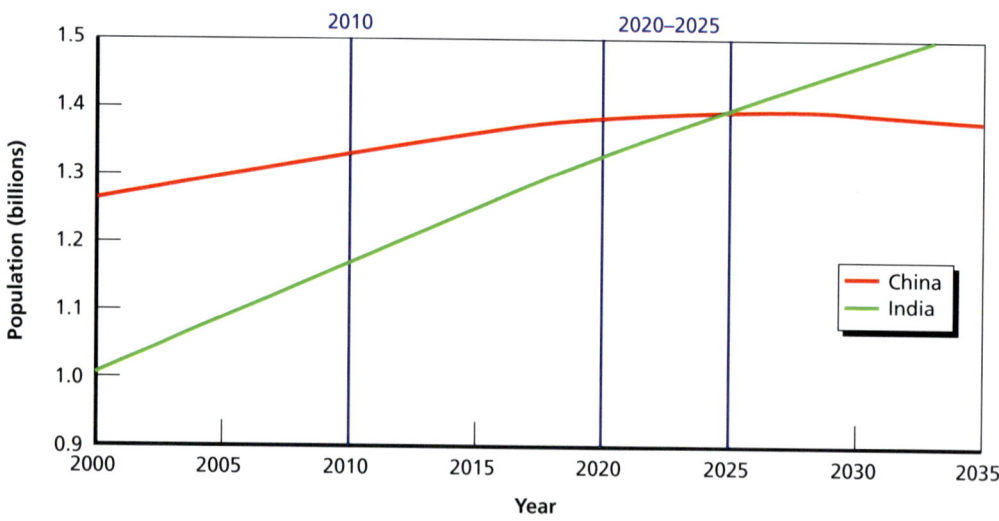

SOURCE: U.S. Census Bureau, 2010.
RAND MG1009-2.1

---

[3]   2050 is the latest year for which IDB shows projections on its website.

CBR of 21.34 per 1,000, but only 8.83 million deaths, resulting in a CDR of 7.53 per 1,000.

Components of national population growth in 2010, including net immigration, are shown in Table 2.1. The IDB estimates that China's population will increase by 0.49 percent in 2010, while India's will increase by 1.38 percent. (By comparison, the U.S. population—bolstered by net immigration of 4.25 per 1,000 population—is expected to grow 0.97 percent in 2010.) Table 2.1 shows that India's higher rate of population growth is largely due to its considerably higher CBR, though the difference in the net immigration rate also contributes modestly. We will now examine the trends in births and deaths in the two countries.

Birth Rates

The CBR in India is projected to exceed that in China over the entire 2000–2035 period (Figure 2.2). During the 2020–2025 period, the CBR difference between the two countries is projected to be slightly smaller (7.2 to 7.4 births per 1,000 population difference) than what it is in 2010 (9.2 per 1,000), and it will shrink to 6.4 per 1,000 by 2035.

During the 2000–2035 period, the CBR in India is projected to decrease smoothly, from 26.0 per 1,000 to 15.0 per 1,000. In contrast, the CBR in China fell from 12.9 to 11.4 between 2000 and 2006 but is expected to increase to 12.3 in 2011–2012, after which it will decline again, to 8.6 in 2035. The slight CBR increase projected for China between 2006 and 2011 is an "echo effect" of the post–Cultural Revolution baby boom; i.e., the women born during that period are now having babies.

The total fertility rate (TFR), the average number of lifetime births per woman, is a measure of fertility that is not affected by the number of women of childbearing age in the population, and the TFR is thus considered a better measure than the CBR for comparing fertility levels between countries or time periods. China's TFR has been lower than India's for many years (Figure 2.3). The IDB estimates that in 2010 the TFR in India is 2.65 children per woman, while in China it is 1.54; i.e., each Chinese woman is currently having, over the course of her lifetime, an average of more than one

Table 2.1
Components of Population Change, China and India, 2010

Demographic Rate	China	India
Crude birth rate (per 1,000 population)	12.17	21.34
Crude death rate (per 1,000 population)	6.89	7.53
Natural population growth (per 1,000 population)	5.28	13.81
Net immigration (per 1,000 population)	−0.34	−0.05
Annual rate of population growth (%)	0.49	1.38

**Figure 2.2
Crude Birth Rates, China and India, 2000–2035**

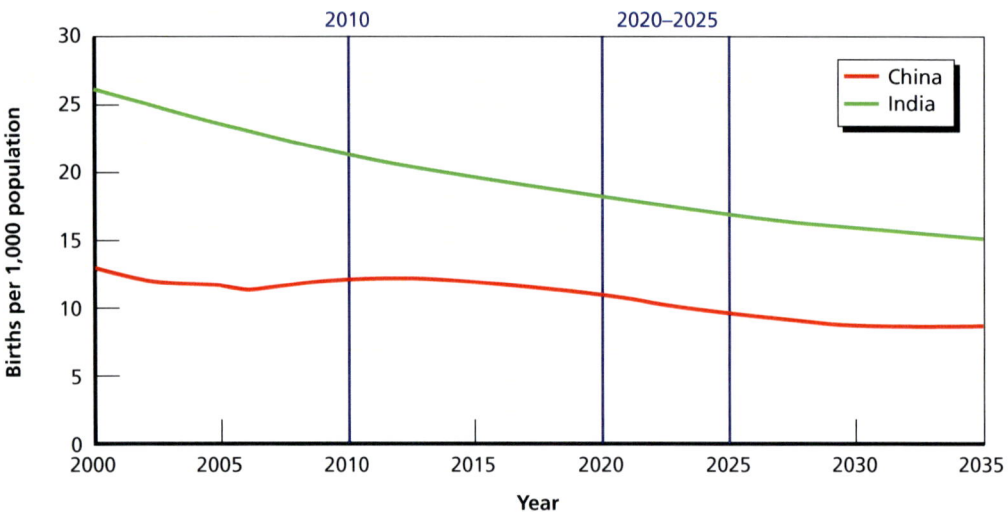

SOURCE: U.S. Census Bureau, 2010.

**Figure 2.3
Total Fertility Rates, China and India, 2000–2035**

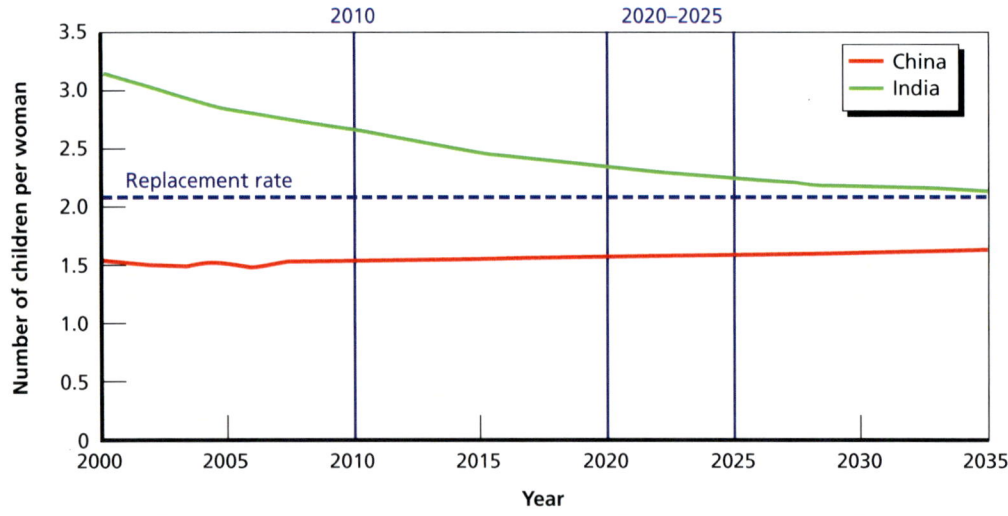

SOURCE: U.S. Census Bureau, 2010.

fewer child than each Indian woman is. The IDB estimates that the TFR in India will decrease very gradually to "replacement level"—the level needed for population stabilization in the long run (approximately 2.1 children per woman)—by 2035.

By contrast, the TFR in China has been below replacement level since 1991.[4] The IDB estimates that the TFR in China decreased to 1.5 children per woman in 2003 but projects that it will start to increase toward 1.6 as we approach 2035. As a result, throughout the period we consider, India's TFR remains higher than China's, though the difference between the two countries will decrease over time. By 2025, women in India are projected to average 0.65 children more than those in China; and in 2050 the difference is projected to be 0.45 children.

The number of births in a country depends not only on the number of births per woman of childbearing age but also on the number of women in this age range. The size of the female population that is of childbearing age (ages 15–49) is currently greater in China than in India (Figure 2.4). However, the number of women of childbearing age in India is projected to increase over the entire 2000–2035 period (and

Figure 2.4
Number of Women of Childbearing Age (Ages 15–49), China and India, 2000–2035

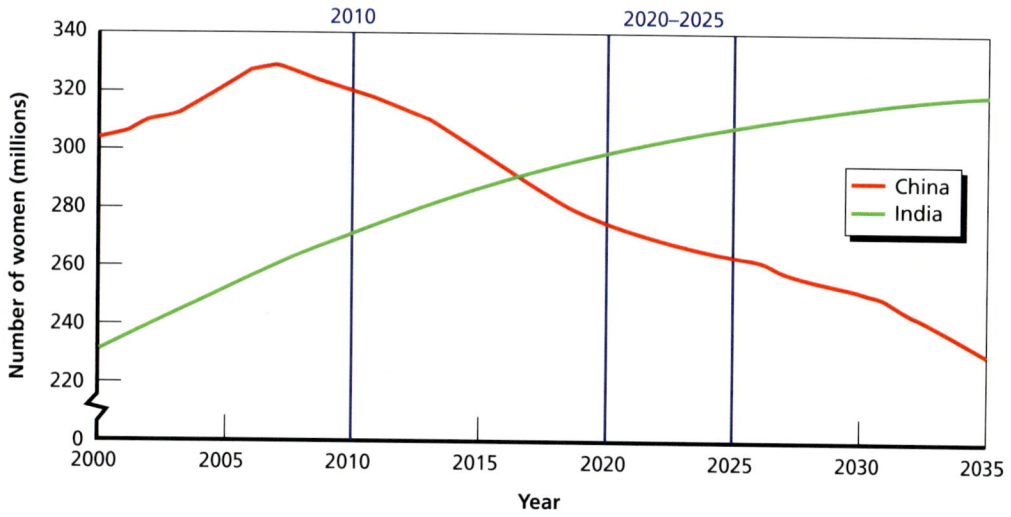

SOURCE: U.S. Census Bureau, 2010.

RAND *MG1009-2.4*

---

[4] Staff at the International Programs Center at the U.S. Census Bureau have told us that the IDB data on fertility are based on the official CBR series released by the National Bureau of Statistics of China in its China Statistical Yearbook, which contains upward adjustments from reported birth data. This official CBR series for 1990–1995 was used by the Census Bureau to generate an implied TFR series based on available age-specific fertility rate patterns and the age structure of women in China for each year of this period. These data correspond closely to the estimates based on new school enrollments, suggesting that 13 percent of children aged 5–9 went unreported in the 2000 census, close to the figure implied by backward projections of those aged 10–14 in the 2005 sample census.

until at least 2050), leading to positive "momentum," while that number has already begun to decrease in China, resulting in negative momentum. India is projected to overtake China in total number of women of childbearing age in 2017. This is why the relative difference in future CBRs shown in Figure 2.2 is considerably greater than that for TFRs in Figure 2.3. (See DaVanzo, Dogo, and Grammich [forthcoming] for more information on fertility trends in China and India, including the role of the one-child policy in China.)

Death Rates

India's CDR is currently higher than China's, and this has been the situation since at least 2000 (Figure 2.5); however, China's CDR is projected to surpass India's in 2014. China's CDR began increasing in 2006 and is projected to continue doing so at an increasing rate over the period of interest. India's CDR is projected to decrease until 2020–2021, after which it will increase slightly. The CDR difference between the two countries is projected to grow throughout the 2000–2035 period, leading to increasingly lower population growth rates in China relative to India.

The CDR is strongly affected by the age composition of a population. Indeed, one reason why CDRs will be higher in China than in India after 2013 is that China's population is, on average, older, and older people are more likely to die than younger people. A better measure for comparing mortality risks or overall health between countries and across time periods is life expectancy at birth (LEB)—the number of years that a person born in a given year can expect to live if the age-specific mortality rates

Figure 2.5
Crude Death Rates, China and India, 2000–2035

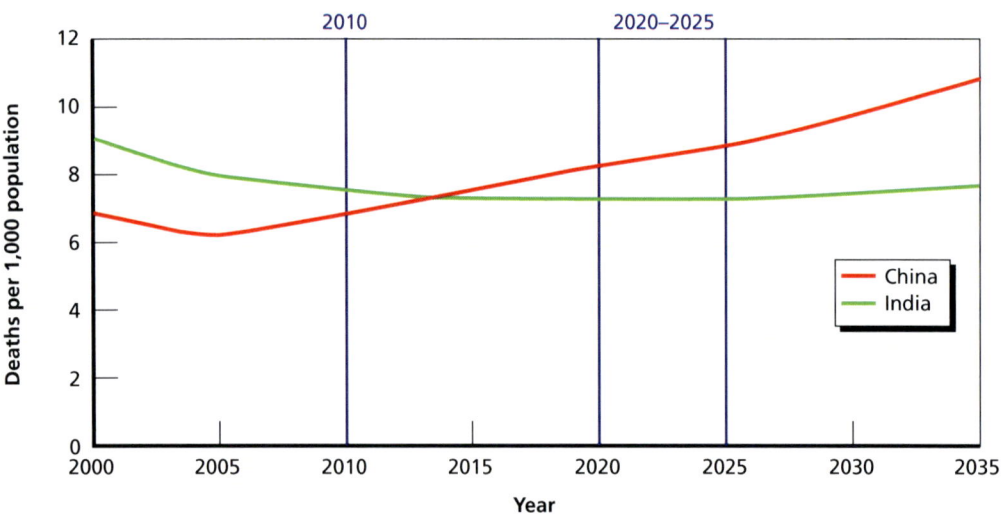

SOURCE: U.S. Census Bureau, 2010.
RAND *MG1009-2.5*

of that year apply throughout that person's life. LEB has been increasing in both countries and is expected to continue to do so through at least 2035 (Figure 2.6). LEB is currently higher in China (74.5 years) than in India (66.5). Death rates from communicable, maternal, perinatal, and nutritional conditions are higher in India than in China for every single cause (WHO, 1999). The LEB gap will narrow somewhat in the future, but LEB in India will lag behind China for the foreseeable future. In fact, it will not be until 2038 that India's LEB will equal the LEB in China today (74.5 years).

Population Growth Rates

In both China and India, the gap between births and deaths is narrowing, resulting in slower population growth in both nations, as shown in Figure 2.7. Population growth rates are expected to be lower in China than in India throughout the 2000–2035 period. India's population growth rate has been declining since before 2000 and is expected to do so at about the same rate throughout the period shown. By contrast, although China's population growth rate is considerably lower than India's, China's rate is quite flat between 2002 and 2011 but is expected to fall somewhat more rapidly than India's thereafter. Beginning in 2027, the number of deaths in China is expected to exceed the number of births, resulting in natural population loss.

Figure 2.6
Life Expectancy at Birth, China and India, 2000–2035

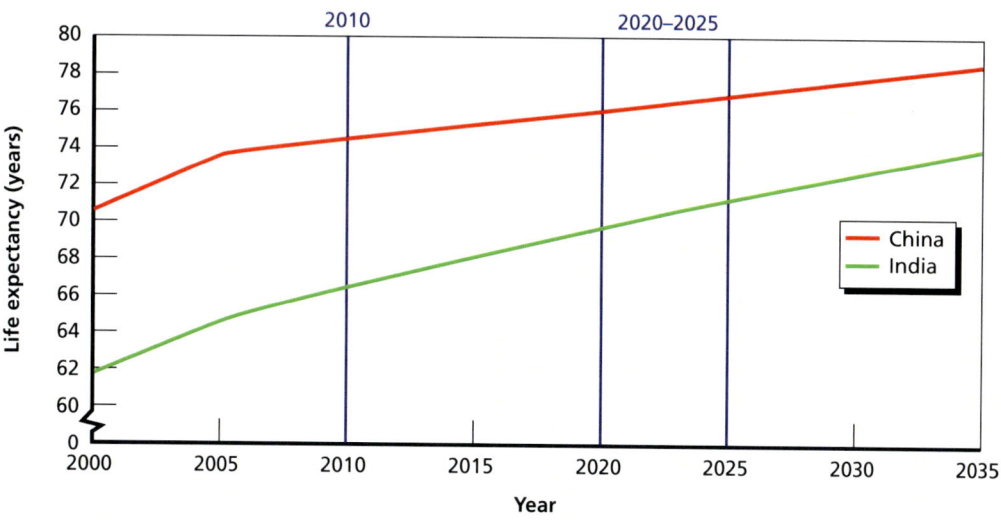

SOURCE: U.S. Census Bureau, 2010.
RAND MG1009-2.6

Figure 2.7
Population Growth Rates, China and India, 2000–2035

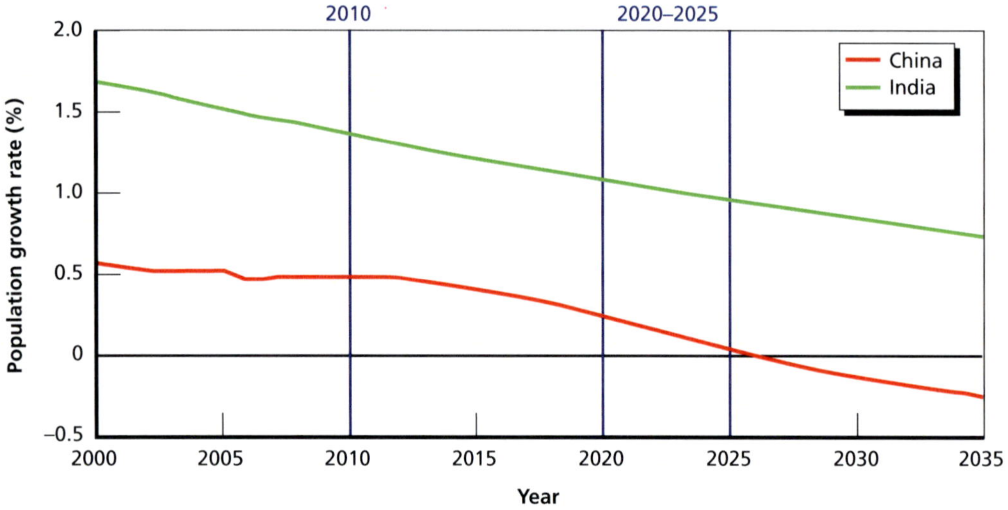

SOURCE: U.S. Census Bureau, 2010.

RAND *MG1009-2.7*

Age-Sex Structure of the Population

The trends in fertility and mortality discussed above affect the present and projected age distribution of a population. Demographers typically use population pyramids to depict the age and sex structure of a population. Such figures are called pyramids because, historically for most nations, particularly in those with persistently high fertility rates, they resemble a pyramid, with a wide base representing large numbers of younger age groups and more narrow bands near the top representing smaller numbers of older people near the end of their natural life span. In Figure 2.8 we show population pyramids for India and China for the years 2000, 2010, 2025, and 2035. The Indian age-sex structure in 2000 is a good example of the classic pyramid shape.

The pyramid shape can still be seen for the 2010 population of India. Because India's fertility rate remains above replacement level and the number of women of childbearing age has been increasing, each birth cohort is larger than the one above it in the population pyramid, though the widths of the "steps" between adjacent bands are smaller for the most recent birth cohorts.

As we go forward, the base (ages 0–4) of the pyramid for India in 2025 is not quite as wide as was it for 2010 (reflecting fewer births in the later year), but above age 20 the bars are all much wider than they are now for those age groups. In 2035, fertility in India is expected to fall nearly to replacement level, and the number of women who are of childbearing age will level off. As a result, the total number of births will

Figure 2.8
Age-Sex Structure of the Populations of India and China, 2000, 2010, 2025, and 2035

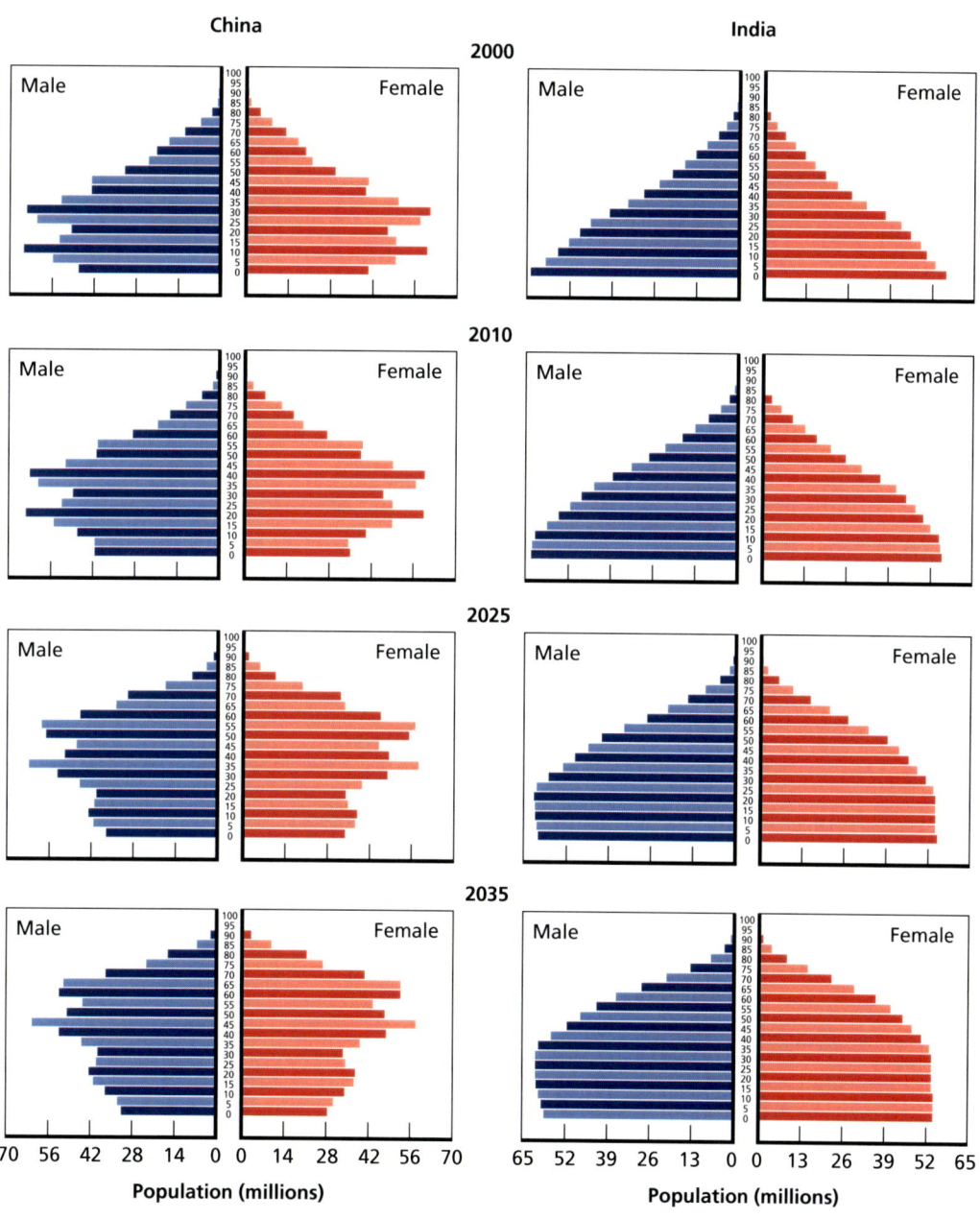

SOURCE: U.S. Census Bureau, 2010.

RAND *MG1009-2.8*

stop increasing, and younger cohorts will become slightly smaller than those immediately older.

The "pyramids" for China have much smaller bases than those for India, especially in future years, and are much more jagged in shape. Each of the pyramids for China has two population bulges. In 2010, there is a bulge for those aged 35–49 (who were born between 1961 and 1975), which reflects the rapid increase in fertility that followed the Great Leap Forward of 1958 and the three subsequent "Black Years" of famine from 1959 to 1961. The smaller cohorts of persons aged 25–34 in 2010 (born between 1976 and 1985) likely reflect China's renewal of family planning campaigns in 1971. The large number of persons aged 15–24 in 2010 (who were born between 1986 and 1995) may reflect legal changes in the marriage age that led to earlier marriages and childbirth, ironically shortly after the introduction of the one-child policy, as well as some population momentum from persons born in the 1960s who then married and had children in the 1980s. The small cohorts born in recent years reflect the low fertility rate following the implementation of the one-child policy.

The "bulges" of the Chinese population pyramid will move upward in coming years as the large cohorts age. For example, in 2035, the large cohort that is now 35–49 will be 60–74. In 2035 there will be many more older people than there are now; for example, there are projected to be 103 million people in China aged 65–69, compared with 40 million in 2010.

Working-Age and Dependent Populations

All these changes in population age composition will affect the percentage of the population that is of working age (typically defined as ages 15–64)—members of the population who *can* (but not necessarily *will*) contribute to the economy—as well as the percentage of population that is of "dependent" age (0–14 and 65+), presumed to be too young or too old to support themselves through labor market activity and who therefore need to be supported by others, typically the family or the state.[5]

Youth

The percentage of the population that is young (under age 15) is projected to be higher in India than in China throughout the 2000–2035 period, though this percentage is projected to decrease steadily in both countries (Figure 2.9). The difference between the countries is currently at its maximum and will be smaller in the 2020–2025 period (average of 8.8 percent) than it is now (12.2 percent).

---

[5]   Not all people aged 15–64 will work, and some who are younger or older may work. Nonetheless, it is generally presumed that, on average, people aged 15–64 produce more than they consume, while the opposite is true for those who are younger and older.

Figure 2.9
Percentages of the Population That Are Under Age 15, of Working Age (15–64), and Older Age (65+), China and India, 2000–2035

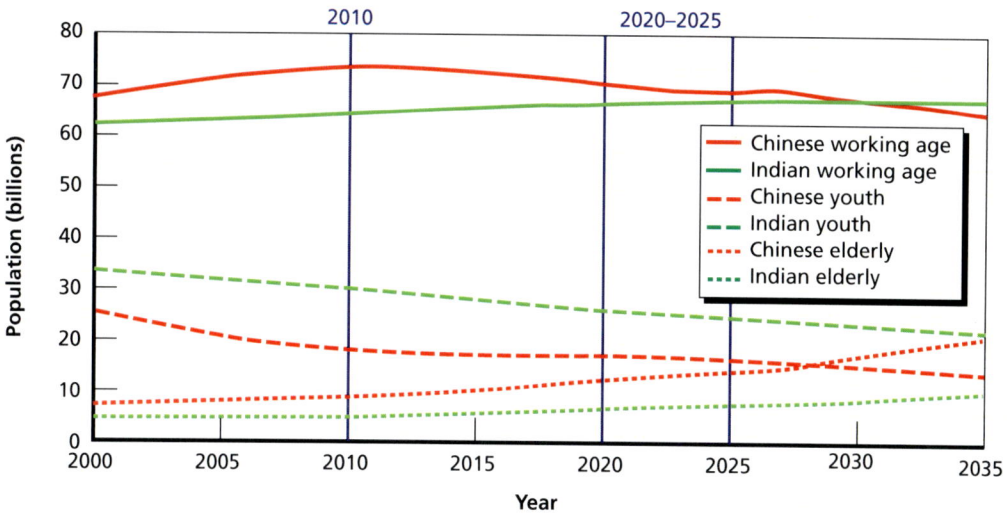

SOURCE: U.S. Census Bureau, 2010.
RAND *MG1009-2.9*

Working-Age Population

China has had a larger percentage of its population that is of working age than India since the mid-1970s (Ahya et al., 2006). Furthermore, as seen in Figure 2.9, in the first decade of this century this percentage was increasing more rapidly in China than in India. This is largely due to the large number of people born in China in the 1960s and early 1970s who were joined in the workforce by their children born in the 1980s and early 1990s (demographic echo effect), as evidenced by the corresponding bulges in the population pyramids shown in Figure 2.8. The percentage of the population of China that is of working age is expected to peak now and to decrease after 2011 (except for a very slight increase in 2026–2027 (the last period when the large post–Great Leap cohorts will still be of working age).

The trend in the percentage of the population that is of working age is more linear in India, with increases in this indicator reflecting steady, nearly linear decreases in the fertility rate. The percentage of the population that is of working age in India is expected to crest around 2030—the same year that India will surpass China on this statistic—and then decline very slowly, reflecting an expectation of decreasing fertility. It is important to note that this decline in India will be very gradual, compared with a much steeper rate of decline in China. (The percentage varies by less than one point in India over the 2019–2035 period, whereas it decreases by 6.2 points in China over the same period.) The difference between the two countries in the percentage of the population that is of working age is currently at its maximum (73.4 percent in China, 64.6 percent in India).

Another indicator of the overall shift in balance between the two countries is that the total number of people of working age in India is projected to overtake that of China in 2028 (when there are projected to be 971 million people of working age in India and 956 million in China [U.S. Census Bureau, 2010]). Furthermore, around this time the working-age population in India will be younger than that in China (Figure 2.10), providing the foundation for growth but also creating a need for entry-level jobs. Meanwhile, there will be more people aged 35–64 (and especially aged 50–64) in China. Nonetheless, it is important to note that throughout our period of focus, 2020–2025, the percentage of the population that is of working age will be larger in China (70.6 percent in 2020, 69.2 percent in 2025) than in India (67.0 percent in 2020, 67.5 percent in 2025), albeit older.

Older Population

In 2010, 5.3 percent of India's population and 8.6 percent of China's is aged 65 or older. In both countries, this percentage will increase, and at an increasing rate (Figure 2.9). By 2025, these numbers will be 7.7 percent in India and 14.3 percent in China, and by 2035 they will be 10.2 percent and 21.0 percent, respectively. By 2035, both China and India will have more than twice as many older people in relative terms as they do now (and an even higher ratio in absolute terms).

Figure 2.10
Age Breakdown of Working-Age Populations in China and India in 2025

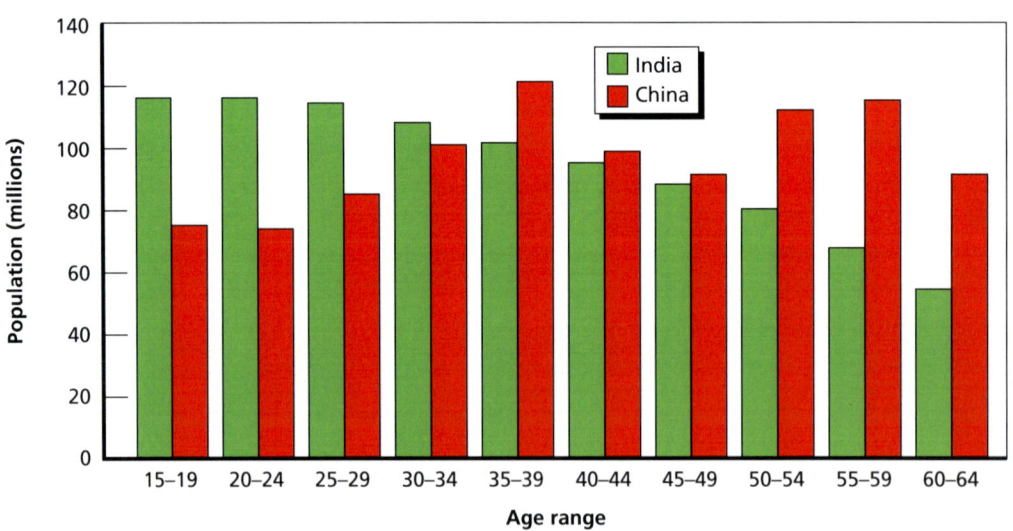

SOURCE: U.S. Census Bureau, 2010.

RAND MG1009-2.10

In terms of the absolute numbers of people, China will have more elders than India in every age subgroup, and they will be relatively older (see Figure 2.11 for data for 2025). Currently, a higher proportion of elders (people aged 65+) is of age 75 or older in China (37.4 percent) than in India (29.9 percent). The difference will narrow considerably by 2025 (when the figures will be 35 percent and 32 percent, respectively). In 2025, both countries will have higher proportions of their elderly who are aged 85 or older (6.6 percent in China and 5.1 percent in India) than they do now (5.8 percent and 3.6 percent, respectively), meaning that the elderly population will be more frail than it is now and thus more difficult to take care of.

Dependency Ratios

The trends in the working-age and dependent populations just presented determine the trends in *dependency ratios*, i.e., ratios of persons of "dependent" ages to those of "working" age. The dependency ratio can be decomposed into the part for youth dependents (under age 15) and that for old dependents (age 65+).

At present in China, there are 36.2 dependents for every 100 persons of working age. Of these dependents, 67 percent are youths and 33 percent are at least age 65. The youth dependency ratio has decreased in recent years, as a result of absolute decreases in the number of persons under 15 years of age (which is a result of a decreas-

Figure 2.11
Age Composition of Older Population (Age 65+) in China and India in 2025

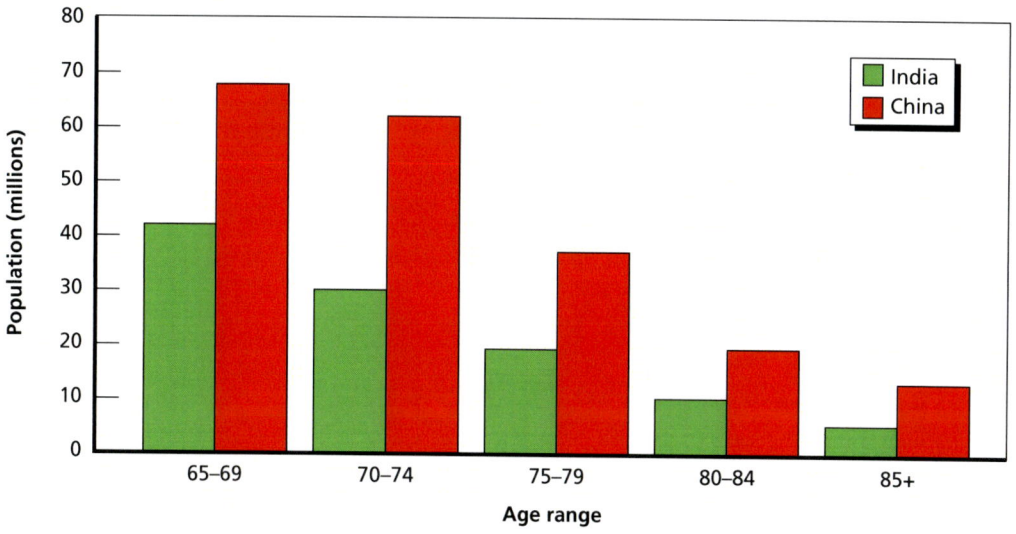

SOURCE: U.S. Census Bureau, 2010.

ing number of births). The youth dependency ratio will increase slightly between 2014 and 2022 (due to the current increase in the natural birth rate) and decline thereafter (Figure 2.12).

The old-age dependency ratio has increased modestly in China in recent years, in part because of improvements in survival to older ages. The ratio is currently 11.8 elderly dependents per 100 persons of working age but is projected to rise rapidly and surpass the youth dependency ratio by 2029, as persons born shortly after the 1949 revolution and before the implementation of family planning programs reach the age of 65. In 2035, there will be 10.8 more elderly dependents than young dependents for every 100 people of working age.

The pattern in dependency ratios is quite different in India. At 46.6 people under age 15 for every 100 of working age, India's 2010 youth dependency ratio is nearly twice that of China's (24.4). India has fewer old-age dependents than China, but the difference between the two countries in their old-age dependency ratios (8.3 for India, 11.8 for China) is much smaller than it is for youth dependents. Currently, 85 percent of India's dependents are youths, compared with 67 percent in China. India's youth dependency ratio has been decreasing slowly in recent years and is projected to continue a slow, steady decrease throughout our study period, to 33.2 in 2035. India's old-age dependency ratio is projected to increase slowly over the study period, but the rate of change will increase over time. Nonetheless, even by 2035 there will be more than twice as many youth dependents as old-age dependents in India.

Figure 2.12
Youth and Old-Age Dependency Ratios, China and India, 2000–2035

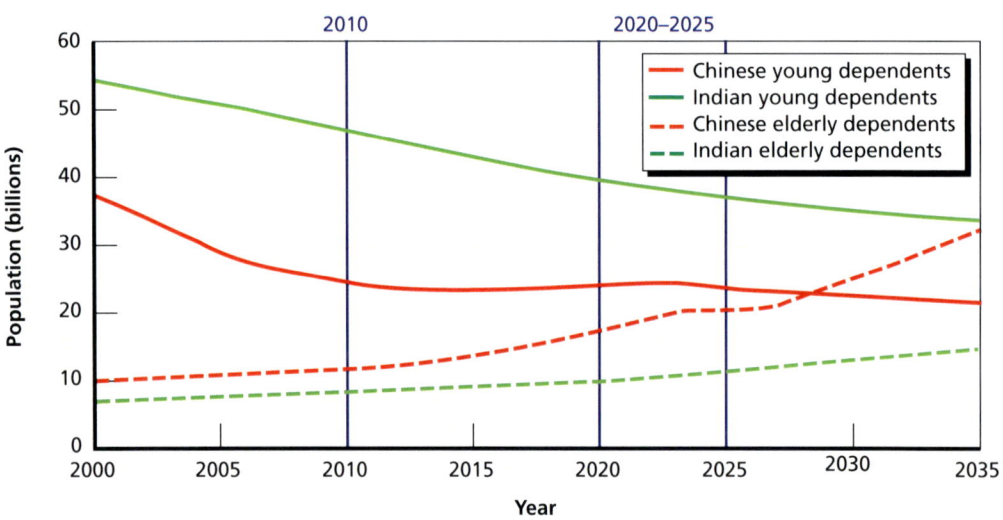

SOURCE: U.S. Census Bureau, 2010.

The overall dependency ratio for China is still falling, but it is expected to begin increasing in 2012 (Figure 2.13), due to continuing increases in the old-age dependency ratio. Throughout the next two decades, India's overall dependency ratio will remain greater than China's; India's overall dependency ratio will fall below China's in 2030. As with every demographic indicator we have considered, changes in India are projected to be less rapid and smoother than those in China.

Sex Ratios

Another demographic trend that may have social implications for China and India is a growing ratio of males to females. This can be seen in the population pyramids in Figure 2.8 (in the wider widths for males than females for most age groups) and in the sex ratios in Figure 2.14. In both China and India, a preference for sons coupled with decreasing fertility has contributed to a higher ratio of males to females among successively younger cohorts (Das Gupta et al., 2003; Lane, 2004; Poston, 2002). Most parents in China or India want to have at least one son. When they decide (or are encouraged) to have fewer children, they sometimes assess the sex of their fetus and abort those shown to be female (Visaria, 2004; Dyson, 2004; Jha et al., 2006). The fact that the ratios of males to females are much larger at all ages in China and India than

Figure 2.13
Total Dependency Ratios in China and India, 2000–2035

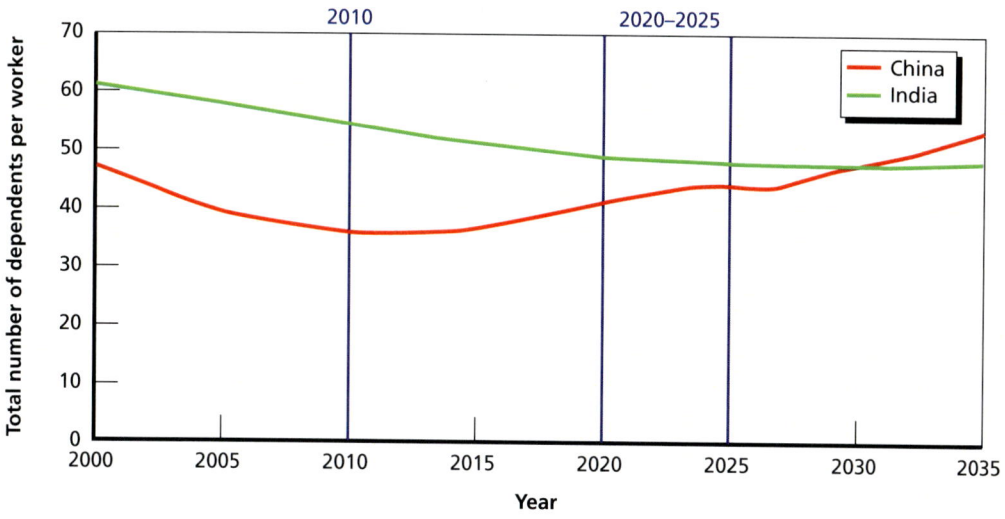

SOURCE: U.S. Census Bureau, 2010.

RAND *MG1009-2.13*

Figure 2.14
Sex Ratios, by Age Group, in China, India, and the United States, 2010

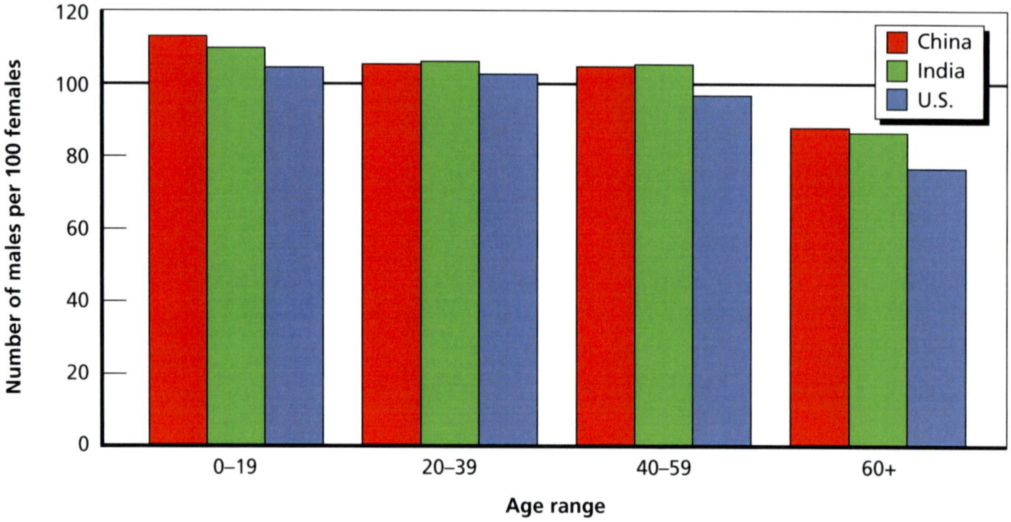

SOURCE: U.S. Census Bureau, 2010.
RAND *MG1009-2.14*

in the United States has been taken as evidence that sex-selective abortion is practiced or that girls are not treated as well as boys, or both.[6]

Opportunity to Reap a Demographic Dividend

An increasing proportion of the population that is of working age provides an *opportunity* to reap a "demographic dividend" (Bloom, Canning, and Sevilla, 2003), through both brute force increase in the numbers of potential workers and an accelerated accumulation of capital due to reduced spending on dependents. Demographic dividends are estimated to have accounted for one-fourth to two-fifths of East Asian per capita GDP growth in late 20th century (Bloom, Canning, and Sevilla, 2003).

The proportion of the population that is of working age will be higher in China than in India until 2030. However, the opportunity to reap a demographic dividend comes from the positive *change* in the proportion of the population that is in the labor

---

[6] It has also been suggested that some of the "missing" girls may exist but are not reported in government censuses and surveys (e.g., Wang and Mason, 2006; Greenhalgh and Winckler, 2005; Dyson, 2001). Nonetheless, estimates that show lower male-to-female ratios than those in official statistics still show higher ratios than are seen in countries such as the United States.

force, which creates the demand and supply signals necessary for economic growth (Bloom, Canning, and Sevilla, 2003). The proportion of the population of China that is of working age is projected to peak during the next two years and then decrease thereafter (except for a flattening in the mid-2020s, which is just before the post–Great Leap baby boom reaches age 65). In India, however, the proportion of the population that is of working age will increase through 2029 and then decrease slowly but steadily afterward. China's demographic window of opportunity is rapidly closing, while India's will remain open until at least 2030 (and changes immediately thereafter will be very small).

Bloom et al. (2009) find that economic growth in China and India between 1980 and 2000 was mainly due to increasing productivity, in large part because of the shift from agriculture to industry and services, but that increases in the proportion of the population that is of working age and in labor force participation rates contributed significantly as well. While the effect of the change in the working-age proportion would, other things the same, favor India, Bloom et al. (2009) find that the *level* of working-age population percentage also matters, favoring China in the years leading up to 2030.

India's working-age population is and will continue to be younger than China's. Younger workers are generally more vigorous and adaptive (Lallemand and Rycx, 2009), and in developing countries they are typically better educated than older workers, but they can be a source of drag if jobs are not available for them. However, an older workforce may not be particularly problematic for China as it tries to develop a post-industrial economy: Recent research (Bloom, Prskawetz, and Lutz, 2007) suggests that productivity may not decline as much with age for more highly cognitive tasks as it does for physical tasks.

Though growth in the working-age population provides an *opportunity* for a country to reap a demographic dividend, the extent to which this occurs depends on the socioeconomic and policy environment (Figure 2.15). Obviously, there must be a demand for the increased supply of labor, and conditions must enable its productivity. This requires effective policies in key areas, including strong health and educational systems to increase the productivity of potential workers; flexibility and competitiveness in the labor market to enable it to absorb the "boom" generations; openness to trade that leads to a growth of productive and rewarding jobs; modern infrastructure and technology to reduce transactions costs and enable economic efficiency; good governance, stable macroeconomics, and a sound financial system to promote savings and investment; and low levels of crime and corruption, which can impede economic progress. Those forecasting the extent of economic growth for China and India note the need for conditions similar to those needed to reap a demographic dividend (Wilson and Purushothaman, 2003; National Intelligence Council, 2004).[7]

---

[7] Both China and India have undertaken reforms, beginning in the late 1960s in China and somewhat later in India, that have increased the roles of markets, opened their economies to international trade, and attracted

Figure 2.15
Linkages and Mechanisms Between Policy and Socioeconomic Influences on Demographics and Economic Growth

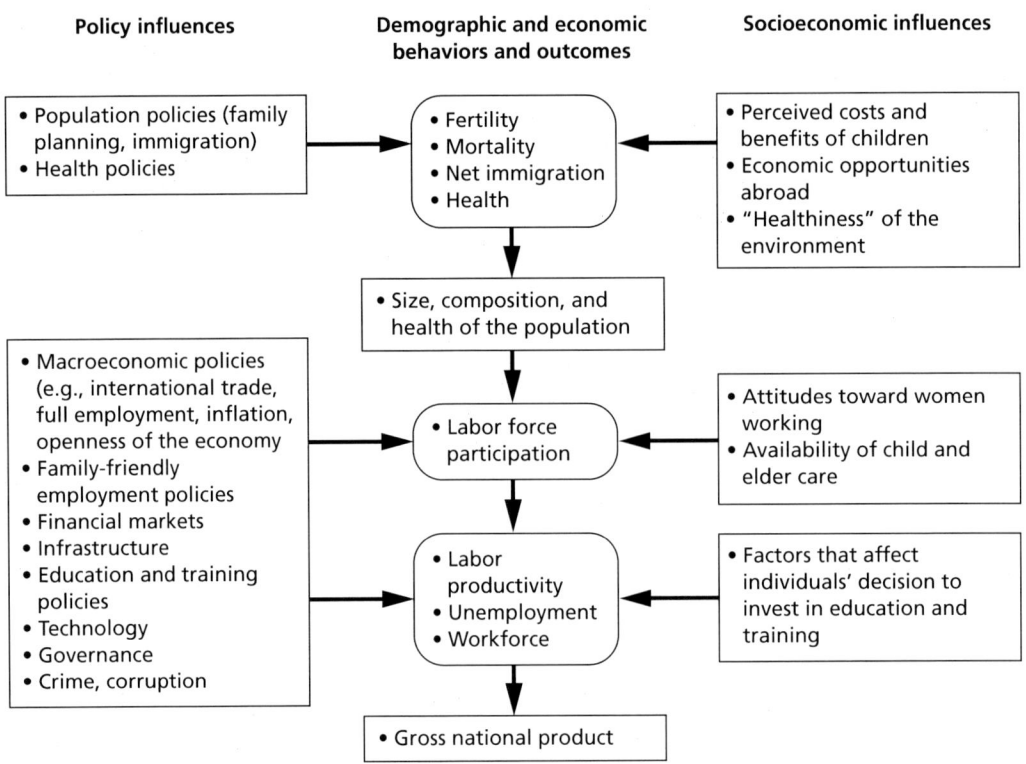

A "second dividend" may arise when the anticipation of population aging incentivizes savings and capital accumulation, but this will not occur if old-age security relies on wealth transfer schemes (Mason and Lee, 2006). Currently, the elderly in both China and India tend to rely overwhelmingly on families for old-age support. In 2006, only 15 percent of Chinese workers were enrolled in pension plans (Cai, 2006), with a net present value of unfunded liabilities exceeding current annual GDP (Eberstadt, 2005). India faces a similar problem; a 2001 study estimated that only 10–11 percent of India's workers had participated in any form of a guaranteed retirement income system (Gillingham and Kanda, 2001).[8]

---

foreign investment. For more on this and the role of sectoral shifts between agriculture and industry, see Bloom et al. (2009) and references therein.

[8]   Gillingham and Kanda (2001) estimate that in 1999 only 11 percent of workers in India participated in any form of a guaranteed retirement income system. Uppal and Sarma (2007) suggest that this number has fallen below 10 percent. The uncertainty in these figures stems from considerable uncertainty about the size of the labor force.

There are no publicly available forecasts that enable us to compare China and India on the presence of all of these conditions. In the sections that follow, we present some recent data on education, health, the role of women, and infrastructure to explore how the two countries differ in key factors that will condition how their demographics will affect their economic prospects.

Education

Large portions of the population that will be of working age in 2020–2025 have already completed their schooling, and those currently in school and recent graduates will be important contributors to this group.

Currently, the population of China is better educated than that of India. In 2002–2003, the most recent year for which data are available, China had higher rates of enrollment at all levels (95 percent vs. 82 percent for net primary enrollment, 70 percent vs. 53 percent for gross secondary enrollment, and 16 percent vs. 12 percent for gross tertiary enrollment)[9] and a considerably higher adult literacy rate (91 percent versus 61 percent) (Goldman, Kumar, and Liu, 2008).

Furthermore, data suggest that the "quality" of schooling is better in China: In 2002–2003, China had a considerably lower average primary pupil-per-teacher ratio than India (17.6 versus 40.2) and spent more per student, especially on secondary education.[10] Improving the quality of education is one of the goals of the 12-year plan announced in China in 2008 ("China Solicits Public Opinions on 12-Year Plan for Education," 2009). By contrast, India is focusing on improving access to education and is just now making investments in such basics as walls, toilets, and running water in schools for all pupils (Ministry of Human Resource Development, 2008). Therefore, the smaller cohorts entering the Chinese labor force between 2020 and 2025 will be better educated than the larger ones in India, placing China at an advantage.

Brain drain from both China and India has been considerable, effectively skimming off the most productive elements of each society. For example, since 1978, more than 70 percent of all the Chinese who traveled abroad to study have not returned home ("China Hit by Brain Drain," 2007), and approximately two-thirds of those

---

[9]   India had a higher rate of gross tertiary enrollment than China prior to 2001, but its growth rate has slowed, whereas China's has grown at an increasing rate. All of the other enrollment rates have been higher in China than in India since before 1985 (Goldman, Kumar, and Liu, 2008).

The gross enrollment rate is calculated by dividing (1) the number of students enrolled a particular level of education, regardless of age, by (2) the population of official school age for that level. The net enrollment rate is calculated by dividing (3) the number of children of official school age for a particular level of education who are enrolled in that level by (4) the total population of children of these ages.

[10]   Based on UN Educational, Scientific, and Cultural Organization (UNESCO) statistics (UNESCO, undated), India has historically spent a higher percentage of its admittedly lower GDP on education than China has, but this trend appears to have reversed in the past ten years. Reliable comparative annual statistics on education are difficult to find after 1999. However, it is known that China has invested heavily in education, particularly with the 2009 stimulus (Bradsher, 2009).

who emigrate from India have some college education (Vonderheid, 2002). It is possible that improving economic conditions may lure some emigrants to return (Dyson and Visaria, 2004).

Health

Health affects not only life expectancy and mortality rates, and hence population growth rates, but also the extent to which working-age people become productive contributors to the economy. Healthy older people may be able to contribute to the economy even after they otherwise might have retired, while improvements in health and longevity can motivate people to save more for retirement (Bloom et al., 2009). The health of a population affects its demand for health care and the resources devoted to it.

Health data for recent years and those projected for the future suggest that the advantage on this dimension goes to China. As of 2004, the average Chinese was healthier than the average Indian (Chatterji et al., 2008). The burden of disease, measured in years of "healthy" life lost, is about 50 percent higher in India (355 years per 1,000 people) than in China (260 years per 1,000 people).[11] As another example, one-sixth (16.6 percent) of respondents in China and nearly half (46.9 percent) in India report having at least one chronic condition (Chatterji et al., 2008). Death rates from communicable, maternal, perinatal, and nutritional conditions are higher for India than for China for every single cause (WHO, undated).

India has a high death rate from infectious and parasitic diseases—many times higher than in China; controlling those will help improvements in life expectancy continue in both countries (see Cook and Dummer, 2004, for China and Horton, 2001, for India). Long-term trends, however, will mainly be affected by how well each nation controls "civilization diseases" resulting from its socioeconomic development. Cardiovascular disease is the leading cause of death in both countries. Cancer accounts for nearly 20 percent of all deaths in China, but only 7 percent in India. Respiratory disease is an important cause of death in India, and to a lesser degree in China (see epidemiology data in He et al., 2005, for China and Joshi et al., 2006, for India). Deaths due to respiratory disease may increase in both countries as pollution increases. By one estimate (Havely, 2005), China has seven of the ten most polluted cities in the world. The World Health Organization (WHO) rates New Delhi as the fourth most polluted city in the world in terms of suspended particulate matter (Gadhok, n.d.).

HIV/AIDS is another factor that will affect trends in life expectancy. Mortality from HIV/AIDS was estimated to be much higher in India (18 per 100,000 population) than in China (0.6 per 100,000) in the late 1990s, but the disease is expanding

---

[11] Technically, these are called the "disability-adjusted life years," or DALYs. For a more detailed discussion of health measurements and scores in India and China, see data presented in DaVanzo, Dogo, and Grammich (forthcoming).

rapidly in both nations (UNAIDS, 2006a, 2006b). In 2007 (the most recent year for which data are available), India was estimated to have 2.31 million persons living with AIDS, the highest total of any country in the world, though its prevalence rate among adults, 0.34 percent, is much lower than the 20 to 30 percent in the most-affected countries in Africa (National AIDS Control Organisation, 2007). The corresponding numbers are much smaller for China: 650,000 persons living with AIDS and a prevalence rate of 0.1 percent, although this is believed to be increasing rapidly (Gordon, 2002). In both nations, AIDS is geographically concentrated, primarily in southern regions and in areas attracting migrants.

Chatterji et al. (2008) project that the number of healthy years lost will decline in China for all broad categories of diseases, indicating that, though older, the population of China will be healthier. As India transitions from communicable to noncommunicable diseases dominating, they project that the number of healthy years lost will *increase* for noncommunicable causes.

Per capita health expenditures nearly doubled in China between 2000 and 2006 (WHO, 2006), and increased by more than 50 percent in India over the same period (WHO, 2009). Furthermore, the WHO expects these costs to keep increasing in the decades to come. Because health care costs increase significantly with age, the burden of the older population will be significantly larger for China than for India, though, as noted above, currently elders are healthier in China than in India.

Women in the Economy

A significant determinant of future economic growth in both countries will be the degree to which women participate in the formal economy (Apps and Rees, 2001; Bloom et al., 2009; Fortin, 2009). In both countries, women are much less likely to participate in the formal labor force than men are, but the difference is much greater in India. In 2006, 69 percent of women in China participated in the formal economy, while in India the rate was only 34 percent (Cook and Chen, 2007).

Attitudes regarding women's roles are presently more permissive in China than in India. For example, 87 percent of Chinese respondents to the World Values Survey feel that a university education is as important for a girl as for a boy, compared with 50 percent in India (World Values Survey Association, 2009; DaVanzo, Dogo, and Grammich, forthcoming). As a consequence, China appears better positioned than India to welcome women into the formal workforce.

Females lag behind males in literacy and educational attainment in both countries, but this is particularly so in India (Goldman, Kumar, and Liu, 2008). In 2000–2001, women in China trailed men in adult literacy by eight percentage points (87 percent versus 95 percent), whereas in India the difference was 25 percentage points (48 percent versus 73 percent). In 2000, less than half of adult women in India were literate. Not only does educating women prepare them to be productive members of the labor force, it also has the additional effects of reducing the number of children desired (which may be useful in India but moot in China because of its one-child policy) and promoting investments in their children's health and education (World Bank, 2009).

Though women in both countries will have fewer responsibilities for child care because of declining fertility, as the populations age, work opportunities for women may be constrained by the need to provide care to elderly parents and parents-in-law, a responsibility they will share with fewer, if any, siblings.

Infrastructure

A well-developed infrastructure can reduce transactions costs, enable economic efficiency, increase the productivity of labor, and alleviate limitations of an aging society by easing movement and extending productivity into later years. The construction of such infrastructure can also provide employment opportunities.

China ranks considerably ahead of India on many dimensions of infrastructure, especially those related to communications and energy. This was not always so; it is a result of a recent, systematic campaign of reinvesting national savings into infrastructure, resulting in rapid growth over the 1980–2005 period. This is a pattern followed by other authoritarian, high-growth economies in Asia, where centrally planned infrastructure is systematically built ahead of demand, promoting export-oriented growth (Akteruzzaman, 2006).

India, by contrast, has taken a less comprehensive and more decentralized approach to infrastructure development, a result of its democratic governance structure, lower GDP, and consistent fiscal deficit. The rate of investment in physical infrastructure as a percentage of GDP has consistently fallen since the early 1990s in India, resulting in an increasing gap between India and China. For example, China's annual investment in its road network increased from about US$1 billion in 1991 to around $38 billion in 2002 (Kim and Nangia, 2008). With over 30,000 km of expressways, China is rapidly catching up with the United States, which has the world's largest road network. In response to the 2008 financial crisis, the Chinese government created a 4 trillion renminbi (approximately US$586 billion) stimulus plan, 38 percent of which went to infrastructure investment. The combination of continued investment and targeted, efficient stimulus funding will result in China adding 5,000 km of expressway every year. By comparison, India's existing national highway network is slow and heavily congested. The story is similar for other infrastructure sectors (Kim and Nangia, 2008). A recent study by McKinsey and Co. (Gupta, Gupta, and Netzer, 2009) concluded that there are severe inefficiencies in the government implementation of infrastructure projects in India and that these may cost India up to 10 percent of potential GDP in 2017–2018.

Other Implications of Changes in Population Age-Sex Composition

Implications of Gender Imbalances and the Changing Composition of "Dependents"

Some have speculated that many of the "excess men" in China and India will not be able to find wives, resulting in a "bachelor bomb" that could lead to social instability and violent crime and foster an authoritarian political system to control perceived increases in violence by such males, lead to larger armies that pursue expansionist policies, or even cause public health problems because of more widespread prostitution (Hudson and den Boer, 2004; Poston and Morrison, 2005). Such arguments, however, may not sufficiently acknowledge the possibility that differences in age at marriage for men and women may increase; that is, men may still marry, but at older ages, while women may marry at younger ages. "Excess" men may also emigrate, or brides may be "imported" from other countries. A recent study of China (Edlund et al., 2007), however, did find that regions with higher sex ratios have higher rates of crime.

The contrasting composition of dependency ratios in each nation indicates that issues related to youth, such as education, will be more prominent in India, while issues regarding the elderly, such as pensions and geriatric health care, will be more prominent in China. It is not clear whether China has dedicated the resources necessary to support larger numbers of the elderly with fewer youths. State-owned enterprises that traditionally funded social programs for workers and their families have largely collapsed, and no national social welfare system has yet replaced them.

Elderly in both India and China traditionally rely on family members to care for them in their old age. With fewer children, parents may expect less support from their families. When the parents of the only children born under China's one-child policy reach old age, they will have, at most, one surviving child to support them, and if that child is a daughter who follows long-standing cultural norms, she may give more attention to her husband's parents than to her own. In 2025, roughly 30 percent of Chinese women at least 60 years of age will have never born a son (Eberstadt, 2005). If the sex imbalance results in an increase in the percentage of men who do not marry, this will mean that many elderly will not have a daughter-in-law to help take care of them. The sex ratios shown above suggest that parents continue to have strong preferences for having sons. It remains to be seen whether these attitudes will change under pressure from demographics. Recent evidence from rural China indicates that the incidence of young couples residing with the wife's parents has been increasing (Shuzhuo and Xiaovi, 2004).

If children migrate to areas of greater economic opportunity and leave their parents behind, the assistance to parents may change from coresidence and other types of nonmonetary support to monetary transfers, which can be used to purchase goods and services that would have been provided by family members. Alternatively, if adult children decide not to migrate or join the modern labor force in order to take care of their parents, opportunities for economic growth may be constrained.[12]

Implications for the Armed Forces

Both countries have, and will continue to have, very large military-age populations (18–49) and are unlikely to face a shortage of people available for military service, per se. However, the underlying social and economic changes may change the internal culture of the people in the military and by extension the militaries themselves.

At present, most of the conscripts in the Chinese People's Liberation Army (PLA) are the only children in their families, and their representation has been increasing with the entry of the one-child policy cohorts into service, from 20.6 percent in 1996 to 52.4 percent in 2006 (Li, 2001). A study by the PLA found that although "one-child" soldiers are no different from their siblinged comrades in certain aspects, such as personality, training records, and service achievement, they do have a higher prevalence of individualism, egocentricity, and risk aversion, underscoring the importance of the "political work" required by the party wing to adequately socialize these soldiers into the military (Finkelstein and Gunnes, 2006, p. 29).

Officer reenlistments for only-child personnel are likely to be affected by greater economic opportunity and the need to provide for elderly parents, potentially reducing the long-term quality of the non-commissioned officer corps. This is part of an overall decrease in the level of prestige enjoyed by the military in a rapidly modernizing Chinese economy, where the armed forces no longer serve as one of only a few ladders for social advancement.

However, the increasing sophistication of the Chinese society brings along a more technologically attuned pool of labor into the military. The ability of the PLA to harness and mobilize that potential will depend on the status of civilian-military relations and the incentives offered by the defense establishment. Improving both of these elements will require additional investments in manpower, further raising the cost of maintaining an advanced military force. The importance of these factors is likely to increase and become more acute during the 2020–2025 time frame.

India's military will not face the same fertility-induced social problems as China's, as India's fertility rate is still above replacement level, and it is the babies born around 2000–2005 who will be the new conscripts in 2020–2025. However, India will face problems similar to China's in officer and highly technical cadre accession and retention, as the broader economy will be more strongly competing with the military for talent. This may even be more acute in India, because the higher levels of income disparity[13] will shrink the available pool of highly qualified candidates, thus increasing

---

[12] For a more extensive treatment of the effects of aging in China, see Banister, Bloom, and Rosenberg (2010).

[13] For a discussion of inequality in India, see Bardhan (2003).

the competition and raising the wages that the military will have to offer in order to attract top talent necessary for a military proficient in the full spectrum of warfare.

Regional Differences

Given the enormous geographic sizes of these two nations, it is not surprising that there is considerable variation by region within them. Within China, dependency ratios are, and are likely to remain, most favorable for economic growth in the more densely populated urban areas in the east. In China's most urbanized provinces, total fertility rates are considerably below replacement level, average health status is much better than in rural areas, and large numbers of migrant laborers from rural areas are helping to sustain economic growth. In contrast, rural areas are aging as working-age people move to cities, leaving the elderly behind. Rural-to-urban migration may be affecting not only care of the elderly in China and India, but also other aspects of family life (e.g., the likelihood of marriage, whether those who marry live together).

Within India, total fertility rates have been below replacement level since the mid-1990s in Kerala and Tamil Nadu, but some northern states still have total fertility rates over four children per woman. These northern states are poor and have weak infrastructure, educational systems, and governance, limiting their ability to absorb the upcoming increase in the working-age population as productive members of the labor force (Acharya, 2003).

AIDS is geographically concentrated in both nations, primarily in southern regions and in areas attracting migrants (UNAIDS, 2006a; National AIDS Control Organisation, 2007).

In both countries, barring a substantial increase in the extent of internal migration, such regional variation means that the evolving demographic conditions that aid or hinder economic growth are likely to affect different regions at different times. To date, India has experienced relatively low rates of internal migration, perhaps because children tend to be educated in the local language of each region and hence often do not have language capabilities to work in other areas. This may change as the country realizes that it may need to overcome regional parochialism in order to promote economic growth.

Since 1990, China has had a higher proportion of its population living in urban areas than India, and China is continuing to urbanize more rapidly than India (Figure 2.16). The United Nations Population Division estimates that in 2007 China's population was 42.0 percent urban while India's was only 28.7 percent urban; it projects that China will have an urban majority by 2020, something India will not reach until well after 2040 (UNPD, 2009a, 2009b).

Great urban population concentrations in China could, arguably, make it easier for China to provide social services as well as to develop the economy with lesser

Figure 2.16
Estimated and Projected Percentage of Population Living in Urban Areas, 1965–2050

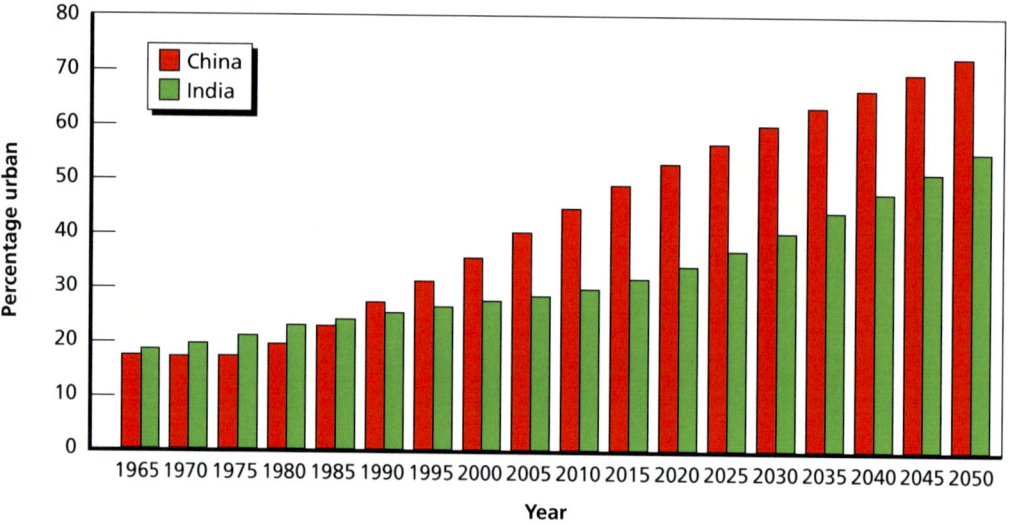

SOURCE: UNPD, 2009b.

RAND *MG1009-2.16*

need of extending national infrastructure to hinterlands. With an estimated 5 million persons leaving farms to move to cities each year in China (Bergsten et al., 2006), however, there are questions about the ability of the labor market to absorb them and whether they will overwhelm existing infrastructure. Further analysis is needed of regional variations and of the potential for internal migration to lessen (or exacerbate) regional inequalities.

Uncertainties and Alternative Scenarios

The data we have presented in this chapter are projections of future demographics in China and India (and estimates of demographics in the recent past). We have used what we believe are the best data currently available, but it is important to keep in mind that these data are based on assumptions about the future (and the past) which may or may not prove to be correct (for example, regarding the impact of HIV/AIDS in the future). In particular, there has long been debate about fertility levels in China, due largely to questions about the extent of birth and child underreporting. Because the impacts of assumptions and uncertainties on estimates of the population size and distribution can be profound, we are implored to mention them.

Assumptions about demographics and uncertainties about the future have led to a wide range of population change–based scenarios for the development of the Chinese economy and its broader society. On one side are the optimists, such as Robert Fogel, whose recent work discusses the possibility of the Chinese economy reaching $123 trillion in annual GDP by 2040, up from $4.9 trillion in 2009 (Fogel, 2010). On the other side of the argument is, for example, Nicholas Eberstadt, who foresees profound negative economic impacts of population aging, gender imbalances, reversal of the rural-to-urban migration flows, and the emergence of a new, unfamiliar family structure based on the changing dependency ratios (Eberstadt, 2004).

Fundamentally, the difference in these projections is based on the expectations of a rapid paradigm shift in Chinese population. For example, Eberstadt's "Perfect Storm" scenario "posits very low fertility in 2040, extremely high sex ratios at birth, a 30 percent decline in both urban and rural marriage rates, and a tripling of divorce rates—and assumes sudden and imminent rather than gradual distant shifts in all these tendencies" (Eberstadt, 2004, p. 17). By contrast, Fogel focuses on the effects of a rapid increase in education, improved rural productivity, and an exceptional ability of the Chinese state to implement population-enhancing policies.

We believe that our focus on more near-term developments between 2020 and 2025 and on the most recent data available allows us to present a balance between these more speculative alternatives. Though demographic changes will certainly pose considerable challenges and provide significant opportunities to both nations in this period, we do not expect those to be of either catastrophic or euphoric nature. While complex systems, such as human populations, are occasionally prone to rapid changes in behavior, often precipitated by unexpected phenomena, such as the HIV/AIDS epidemic, it is important to keep in mind that the entirety of the workforce and the elderly of the 2020–2025 period have already been born; the largest uncertainty concerns the size of the youth population and the resultant socioeconomic impact of changes in the youth dependency ratio.

Summary Assessment of the Relative Strengths and Weaknesses Brought About by Demographic Trends in China and India

It is projected that China's population will remain larger than India's during most of the 2020–2025 assessment period, but that India will surpass China in population size in 2025. China is likely to continue to have higher GDP per capita than India,[14] which matters more on the world stage than numbers of people. In both countries, increasing populations, together with increasing income and affluence, will increase demands on world resources and place strains on the environment (World Bank, 2007; "India's Pollution Crisis," 2008). China has since the mid-1970s had a larger percentage of its population of working age than India; this difference is projected to persist through our period of interest, 2020–2025, until 2030. However, the percentage of China's

---

[14] For extended discussion, see the macroeconomic forecast in Chapter Three.

population that is of working age will peak in the next two years and decline thereafter, while this percentage will be increasing in India. Because more women participate in the labor force in China than in India, the crossover point for the proportion of women that actually works may occur somewhat later.

The opportunity to reap a "demographic dividend" is limited in time. Eventually, the working-age population decreases in relative size, as its retiring members are replaced by smaller cohorts resulting from lower fertility rates. While China has two decades before its overall dependency ratio is projected to exceed India's, it has only one more year before the population of older people starts increasing more rapidly.

When compared with India, in the short term China seems to have more of the preconditions to take advantage of its demographic window of opportunity and to deal with demographics when they become a potential drag: more flexible labor markets; higher rates of female labor force participation, more highly educated women, and more open attitudes about women working; less illiteracy in general (and especially for women); better infrastructure; more internal migration (though much of it "illegal"); and a higher degree of urbanization, more openness to foreign trade, and slightly higher rates of coverage by public pensions. It is for these reasons that we feel that, on balance, China will remain "ahead" of India during the 2020–2025 assessment period.

In the long term, however, China's prospects may be hindered by its demographics. An aging population without an established safety net will create demands for new types of services (particularly health care), reducing the disposable income of the working population through wealth transfers to the elderly and laying claim to the large national savings pool that China has built up during the boom years.

It is our assessment that, on the whole, China's projected demographics are creating a challenge for its economic development—a potential demographic drag—that may be more complex to manage compared with the situation of India. While China was very successful in controlling the size of its population through antinatalist policies in the late 1970s and early 1980s, it is unclear whether it can successfully implement pronatalist policies to avoid a long-term decrease in its population. Not only has the social environment changed, but the goals of these policies are more difficult to achieve—both democratic and authoritarian regimes in Europe found that such policies mostly result in changing in the timing of childbearing rather than the overall number of children (Hugo, 2000; David, 1982; Grant et al., 2004). Furthermore, even if pronatalist policies are successful, it takes around two decades for the babies they produce to become old enough to enter the labor force; in the meantime, the result is an increase in the number of young dependents. India is perhaps facing a more straightforward task, since its primary challenges are improving infrastructure, health care, education, and the role of women rather than altering the behavior of individuals. However, China has a good head start on development and, given its centralized decisionmaking governance structure, will have an easier time implementing socioeconomic policies required for change, but the methods by which it could successfully increase fertility are not obvious.

In the future, India will have more favorable demographics than China, but whether it is able to reap a demographic dividend will depend on successful government implementation of an ambitious economic development agenda. Improving infrastructure, health, education, and the role of women while maintaining social peace in a society that is increasingly stratifying by income requires national consensus with a long-term outlook. Whether such a course is possible in a large, diverse parliamentary democracy such as India is difficult to predict. China's experiences indicate that such policies are feasible, but direct comparison between the two remains difficult.

CHAPTER THREE

China-India: A Macroeconomic Assessment

Introduction: Forecasts of Economic Growth in China and India

Economic growth in China and India has become a particular focus of attention in Asia, in the Asia-Pacific Economic Cooperation (APEC) forum, in the G-20, and in the global economy. In recent decades, growth in both countries has exceeded expectations. Between 1980 and 2008, China recorded an average annual growth in GDP of 9 percent, while India's growth during the same period was about 6 percent. Both countries face the challenge of sustaining such high rates of growth. This chapter summarizes a meta-analysis of growth estimates for China and India for the period through 2025 by three sources: academic scholars, business organizations, and international financial institutions. We also summarize and evaluate the key assumptions underlying the estimates. Finally, we compare five different scenarios of high, low, and average estimates between the two countries, and conclude with several observations based on the meta-analysis.

We conducted a meta-analysis of 27 studies made between 2000 and 2008 of China's and India's recent and prospective levels and growth rates of GDP, capital, employment, and total factor productivity. The studies used in the meta-analysis were culled from a larger set of 47 based on the sufficiency of their data for conducting the analysis. Most of the studies contained explicit projections of these macroeconomic indicators through 2025. Where estimates of the two countries' macroeconomic indicators were implicit, we derived their implicit values using either an incremental capital-output ratio method or a Cobb-Douglas production function. We detail the steps of this meta-analysis in Appendix A.

The analytic methods used in the 27 documents vary widely. Some rely on simple extrapolation and trend analysis in forecasting growth of GDP and its components, while other studies used more sophisticated models. Some analyses concentrate on a single aspect of economic growth—for example, the role of capital accumulation and its determinants in explaining differences between the two economies—while other studies consider several factors and policies affecting growth in the two countries. Similarly, some projections go only through 2020, while others extend through 2050. A few of the studies focus on either China or India rather than on a comparative analysis of both countries. In these few cases, we compare their respective results, while allowing and adjusting for the studies' differing data sources and methods.

Our review of forecasted GDP growth rates in China and India through 2025 suggests that the recent rapid growth of these countries may be sustainable in the future, but at a somewhat slower pace. As shown in Table 3.1 and Figure 3.1, for the entire set of studies, growth rates are projected, on average, to be 5.7 percent in China

Table 3.1
China-India Macroeconomic Meta-Analysis: Summary of Salient Estimates, 2020–2025

	GDP		TFP		Employment		Capital	
	China	India	China	India	China	India	China	India
Mean	5.7	5.6	3.4	2.1	0.4	1.6	6.1	6.9
Max	9.0	8.4	5.6	3.6	0.6	1.9	9.4	9.8
Min	3.8	2.8	2.1	0.N1	−0.1	0.7	4.2	3.9
Variance	2.2	2.3	1.0	1.0	0.0	0.1	2.1	2.5
N (obs)	28	26	28	26	28	26	28	26
N (studies)				27				

NOTES: Growth rates are given in percentage per year. The number of observations does not match the total number of studies because some studies provide estimates for either China or India but not both. TFP = total factor productivity.

Figure 3.1
China-India Macroeconomic Comparisons: Salient Estimates, 2025

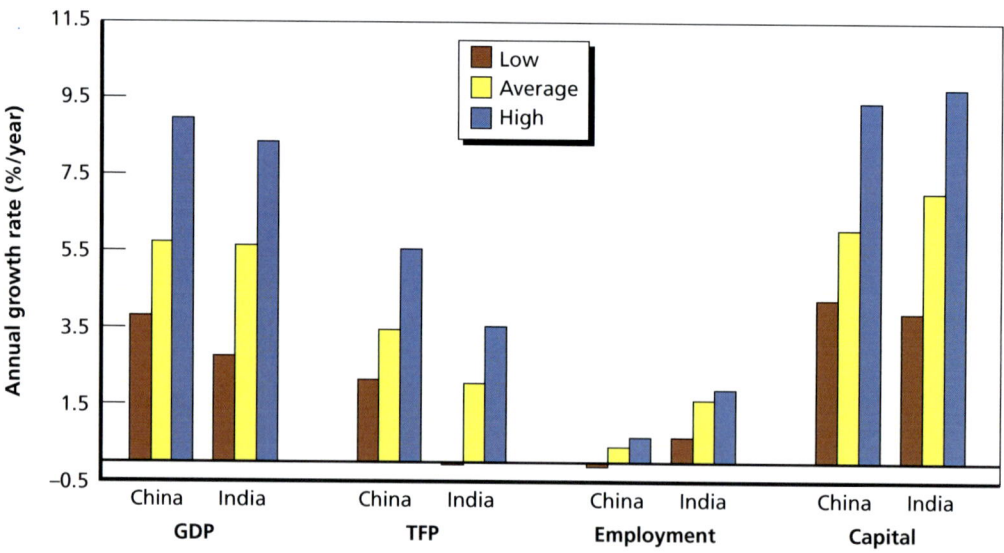

and 5.6 percent in India during the period from 2020 through 2025. During the same period, the average estimates for growth in the accumulated stock of capital are 6.1 percent for China and 6.9 percent for India; for growth in employment, the average estimates are 0.4 percent for China and 1.6 percent for India; and for total factor productivity, the average estimates are 3.4 percent for China and 2.1 percent for India.

These estimates of average rates are, unsurprisingly, accompanied by major uncertainties, as suggested by the wide range between the highest and lowest growth estimates, as well as their corresponding variances.[1] In the 27 studies we selected, the estimates for GDP growth rates for the 2020–2025 period range from 3.8 to 9.0 percent for China and from 2.8 to 8.4 percent for India. For the growth rates in capital stock, estimates range from 4.2 to 9.4 percent for China and from 3.9 to 9.8 percent for India. For employment growth rates, the ranges are from −0.1 to 0.6 percent for China and from 0.7 to 1.9 percent for India. And for total factor productivity, the estimated growth rates range from 2.1 to 5.6 percent for China and from 0.1 to 3.6 percent for India.[2]

In the following section, we review these estimates by separating them into three clusters: (a) those by academic authors and institutions, (b) those by business organizations and authors (e.g., Goldman Sachs, PricewaterhouseCoopers, McKinsey), and (c) those by international organizations (e.g., the World Bank, the International Monetary Fund). We also evaluate the differing assumptions that explain the wide range of the estimates and highlight several other contrasting aspects of the three clusters.

Studies by Academic Authors and Institutions

We included 11 studies by academic authors and institutions in our set of 27 studies.[3] The academic cluster of studies generates the widest range of growth estimates for both China and India; the range of estimates for this cluster is summarized in Table 3.2. Within this cluster, the highest estimate for China's annual GDP growth for 2020–2025 is 9 percent, and the lowest is 3.8 percent. For India, the estimates for annual GDP growth within this cluster range from 2.8 percent to 7.2 percent. Again, the variances of estimates in this group are higher than for the studies by business groups and international organizations. That the variance in the academic authors' estimates is by

---

[1] We deliberately focus on the extremes of the range, rather than on maximum-likelihood estimates, in order to highlight the uncertainties involved in these forecasts. In turn, later in this chapter we use the maximum and minimum points of the range to formulate high- and low-growth scenarios to compare Indian and Chinese GDP forecasts for 2025.

[2] The corresponding variances will be discussed below.

[3] The 11 in the academic cluster are Brown (2005), Golley and Tyers (2006), Hofman and Kuijs (2007), Holz (2005), Huang (2003), Laurent (2006), Linn (2006), Paltsev and Reilly (2007a), Paltsev and Reilly (2007b), Poncet (2006), and Tyers et al. (2006).

Table 3.2
China-India GDP Growth Estimates by
2020–2025, by Academic Authors (%/year)

	China	India
Low	3.8	2.8
High	9.0	7.2
Average	5.5	4.3
Variance	2.6	2.4
N (observations)	15	9
N (studies)	11	7

NOTE: Differences between the numbers of observations and of studies arises because several studies included both high- and low-growth forecasts by their authors.

far the largest among the three clusters is a finding that will be surprising to some readers, but unsurprising to others!

In making these estimates, the academic authors tend to focus especially on one or two particular aspects of each country's growth. For example, some studies in this cluster focus especially on demographic changes in China and India (Golley and Tyers, 2006; Tyers et al., 2006). Another forecast in this cluster focuses especially on the two countries' roles in global energy markets, their greenhouse gas emissions, and the resulting effects on growth (Paltsev and Reilly, 2007a). Still another study in this cluster focuses heavily on China's regional economic structure, basing its estimates on sectoral and regional economic growth (Huang et al., 2003). Several studies base their long-term forecasts on a standard Cobb-Douglas production function (Holz, 2005; Poncet, 2006).

Holz (2005), from the Hong Kong University of Science and Technology, projects the highest annual GDP growth rate for China (9 percent) for the 2020–2025 period, arguing that China can expect many years of rapid economic growth. Holz's forecast mainly rests on projecting recent growth rates into the future; he concludes that China's GDP will surpass that of the United States in purchasing-power terms by the middle of the next decade.[4] He suggests this trajectory would follow the examples of Japan, Taiwan, and Korea in the early stages of their development. Holz contends that the structural changes taking place in China, along with factor price equalizations, will match the patterns of growth achieved by these other countries.

---

[4]   Holz assumes that the U.S. average annual GDP growth rate during this period is going to be 3.0 percent.

In sharp contrast, another author in this cluster (Laurent, 2006) suggests that China's average annual GDP growth rate in 2020–2025 will decline to 3.8 percent, reasoning that declining numbers of prime working-age workers will inhibit China's growth. Laurent contends that India's growing labor force might enable India to grow more rapidly if its populous were more highly educated. However, unlike other economists who compare the two economies, Laurent is pessimistic about India's ability to educate its population. He expresses little confidence that this will happen, and forecasts that India's average annual GDP growth in 2020–2025 will fall to 3.4 percent.

The effect of declining population on economic growth is also among the most significant issues raised in the study by Tyers et al. (2006) from the Australian National University (ANU). Their study suggests that relative labor abundance in India will bring higher capital returns and attract a rising share of global foreign direct investment to India.[5] Accordingly, the Tyers et al. forecast for India's annual growth rate in 2020–2025 is 7.2 percent. The authors believe that India will displace China as the world's most rapidly expanding economy.

The ANU authors also examine the economic impact of a hypothetical increase in fertility in China that might occur if (1) China were to abandon its one-child policy and (2) a more-rapid-than-expected reduction of fertility were to occur in India. The ANU authors question the plausibility of this scenario by noting that, even if China were to abandon its one-child policy, fertility might not rise substantially; the authors point to the reduction in fertility in China that occurred before introduction of the one-child policy in 1979 (Carnell, 2000) to show that forces other than policy have influenced China's fertility rate. For example, fertility is sensitive to cultural norms as well as economic incentives. Consequently, if a norm of low fertility has indeed taken root in China, it may be difficult to reverse. In India, too, fertility has been falling slowly, as discussed in Chapter Two. An acceleration of fertility reduction in India might occur either as a consequence of economic development or because of exogenous societal reasons.

A contrasting study from scholars at the Massachusetts Institute of Technology (Paltsev and Reilly, 2007b) forecasts that India's average annual GDP growth rate in 2020–2025 may be as low as 2.8 percent, due especially to high energy prices which would put a heavier burden on India's increasing oil imports. However, the MIT paper ignores the possible effects of new technologies that might partly reduce dependence on fossil fuel imports as well as lower their prices.

Similar to Holz (2005), researchers at CEPII, a French research center on international economics (Poncet, 2006), use a production function based on a neoclassical model in which GDP growth depends on growth of the labor force, growth of capital, and growth of total factor productivity. Poncet's GDP projections for both China and

---

[5] The ANU paper by Tyers and colleagues assumes that, with increasingly open capital accounts, both China and India stand to attract foreign investment the more rapidly their labor forces grow.

India are modest, placing them at average annual rates of 4.6 and 4.5 percent, respectively, during the period through 2025. Unlike other studies in the three clusters of our meta-analysis, in which total factor productivity enters the growth forecasts exogenously (Poddar and Yi, 2007; Hofman and Kujis, 2007; Rodrik and Subramanian, 2004), or in which it is modeled as a process of "catch-up" (Wilson and Purushothaman, 2003), Poncet's study links the growth of factor productivity to investment in human capital. Growth in total factor productivity becomes an endogenous function of average years of education and the income gap compared with income in the United States. The resulting differences in total factor productivity and in the growth projections for India and China are thereby enlarged. Taking into account expected improvements in education, Poncet projects the average annual total factor productivity growth in the period 2020–2025 in China as 2.5 percent and in India as 1.9 percent. In particular, Poncet projects that China and India's GDPs could grow at yearly average rates of 4.6 and 4.5 percent, respectively, during a period running up to 2025.

One paper in the academic cluster (Hofman and Kuijs, 2007) estimates an average annual GDP growth rate for China of 6.7 percent during the 2020–2025 period, while suggesting that China's recent 9 percent growth rate is likely to be unsustainable. The authors believe that the greatest threat to China's future growth lies in the internal imbalance between aggregate domestic savings and domestic investment that has developed since 2005. They estimate that China's aggregate national savings have come to exceed aggregate domestic investment by 12 percent of GDP. Some researchers have paid less attention to this imbalance, focusing instead on the external imbalances reflected in China's large and continuing current account surpluses—in fact, the two sets of imbalances are exactly equal to one another, because of the basic accounting identity that defines how the external and internal flow of funds is calculated.[6]

According to Hofman and Kuijs, China's aggregate surplus of savings is due largely to rapid increases in enterprise saving, whereas household and government saving has been stable or declining in recent years (Hofman and Kuijs, 2007).[7] China's corporate sector has been enjoying high profits, while the wage share of GDP has been declining. The authors argue that this disparity is at the heart of China's growing income inequality, and they suggest this is further exacerbated by the low returns earned by China's savers in financial markets (about 2.5 percent), despite the economy's rapid growth.

---

[6] The accounting identity specifies that the difference between an economy's aggregate savings and aggregate investment is equal to the difference between (1) the sum of its exports and other current international earnings and (2) its imports and other current international payments. Thus, the internal and external imbalances are exactly equal.

[7] One author of the current monograph (Wolf) considers this assertion to be erroneous. In fact, household savings (and household holdings of liquid savings balances in the major state banks) have grown substantially in recent years.

Studies by Business Organizations and Authors

We included nine studies by business organizations and authors in our set of 27 studies.[8] Table 3.3 summarizes the range of GDP growth estimates generated by this cluster. The studies in the business cluster are typically rooted in the neoclassical growth model referred to above, and they generate a relatively narrow range of annual GDP growth estimates for China in the period from 2020 to 2025: between 4.5 and 5.2 percent. In contrast, the range of growth estimates for India is considerably wider: between 5.4 and 8.4 percent. The 8.4 percent estimate is from a Goldman Sachs paper (Wilson and Purushothaman, 2003) that optimistically posits high productivity growth, generally favorable demographic factors, and improvements in educational attainment.

Contrary to the average growth estimates of the academic cluster, the business cluster studies place India's expected GDP growth rates in the 2020–2025 period *above China's*—specifically, an average estimate of GDP growth of 6.3 percent for India versus 4.7 percent for China. These figures compare with the academic cluster's average estimate of 5.4 percent for China and 4.3 percent for India. Furthermore, the variance estimates for the business cluster are substantially lower than for the academic cluster.

Not surprisingly, the papers in the business cluster, especially those sponsored by Goldman Sachs, accord particular importance to the prevailing regulatory environment and the protection of property rights in influencing their growth forecasts. This emphasis is missing in the papers by the academic institutions discussed earlier, and in the international organizations' studies.

Table 3.3
China-India GDP Growth Estimates by Business Organizations and Authors, 2020–2025 (%/year)

	China	India
Low	4.5	5.4
High	5.2	8.4
Average	4.7	6.3
Variance	0.1	1.1
N (observations)	6	9
N (studies)	6	9

---

[8] The nine studies in the business cluster are Ablett et al. (2007), Bergheim (2005), Desai et al. (2007), Hawksworth and Cookson (2008), O'Neill et al. (2005), Poddar and Yi (2007), Purushothaman (2004), Wilson and Purushothaman (2003), and Wilson and Stupnytska (2007).

The studies sponsored by the business organizations also tend to compare their estimates for China and India with those of other Asian economies, in particular South Korea—a characteristic that the business cluster shares with the cluster of studies by international financial institutions.

One of the business organization papers, from Goldman Sachs (Wilson and Purushothaman, 2003), suggests that, if certain conditions prevail in China—namely, macroeconomic stability, high investment rates, and a large labor force—the result will likely make China the world's largest economy by 2041, when China's per capita income is estimated to be US$30,000.[9] According to this study, India's growth rate is likely to remain above 5 percent for several decades, and its GDP will exceed Japan's by 2032, reaching a level of per capita income 35 times its current level yet still significantly lower than China's in 2050.

Wilson and Purushothaman make several simplifying assumptions that indeed cast doubt on their final estimates. For example, they do not consider changing demographic conditions in China, instead making the erroneous assumption that the proportion of the working-age population in China will remain stable. In reality, as discussed in Chapter Two, the percentage of China's working age population is expected to peak in 2010–2012 and to decline thereafter. Furthermore, Wilson and Purushothaman make an unrealistic assumption that the investment rate of economies seeking to catch up with the developed countries will remain very high and constant. It is more realistic to assume that, as India and China's per capita income levels approach those of the developed countries, they will experience lower rates of return on investment, and therefore lower their rates of investment, leading to lower rates of growth in total factor productivity. These reduced rates will tend to converge more closely to those prevailing in more advanced economies. For example, developing countries that previously maintained an investment rate of 25–30 percent of GDP are likely to find these rates converging toward prevailing levels in the Organisation for Economic Co-operation and Development (OECD) countries (about 20 percent) (World Bank, undated), as a result of lower rates of return on new investment.

---

[9]   Wilson and Purushothaman assume that the U.S. annual GDP growth rate during the same period is between 2.1 and 3.1 percent.

Per capita income levels are in market exchange rates but closer to PPP exchange rates. Wilson and Purushothaman assume that, as countries develop, there will be a tendency for their currencies to converge toward PPP rates. PPP exchange rates are calculated as the ratio between a market-basket of goods and services (e.g., consisting of consumer purchases or of the country's GDP as a whole), priced according to the country's own prices and weighted by their corresponding shares, divided by the same weighted market basket of goods and services but instead priced in prevailing U.S. dollar prices. Hence, PPP rates omit the effects of capital transactions, which heavily influence market exchange rates, while market exchange rates omit the effects of nontradable goods and services (e.g., residential property values and domestic household services). In developing countries, PPP exchange rates are predictably higher for domestic currencies (e.g., Indian rupees and Chinese renminbi) than are their market exchange rates.

One of the more pessimistic Goldman Sachs papers (Purushothaman, 2004) places India's annual GDP growth at about 5.7 percent during the 2020–2025 period, reasoning that the two crucial conditions of improved infrastructure and expanded education may be insufficient to keep India on a steady growth path. Another Goldman Sachs paper (Poddar and Yi, 2007) calls attention to certain constraints on doing business in India as potential threats to private enterprise.[10]

Another paper sponsored by the Deutsche Bank (Bergheim, 2005) projects India to grow more rapidly than China in the period to 2020. The Bergheim paper forecasts India's average annual GDP growth at 5.5 percent, compared with projected Chinese GDP growth of 5.2 percent over the same period. The major contributor to this gap, according to Bergheim, is China's slower population growth rate (at 0.8 percent annually, about half that of India's growth rate), an outcome of its one-child policy.

Analysts from the McKinsey Global Institute (2006) suggest that India's likely continuation of its recent rapid growth will result in the tripling of India's average household income over the next two decades. If this trend is sustained, India will become the world's fifth-largest consumer economy by 2025, compared with its current position of 12th. Unlike other studies in the business organization cluster, and while noting the progress that India has made to date, the McKinsey paper emphasizes the significant challenges it still faces. These include, for example, the large regional disparities in growth and in poverty levels: For example, India's southern and western states prosper while the northern and eastern states lag far behind. Furthermore, while India has been slowly urbanizing over the past two decades, the McKinsey study sug-

---

[10] A recent World Bank survey of development indicators suggests it costs almost nine times as much to start a business in India as in China, and almost six times longer to close a business in India than in China. These striking results are summarized below.

Business Conditions in China and India, 2007

	India	China	Korea
Starting a business			
Time required (months)	1.1	1.2	0.6
Cost (percent of GDP per capita)	74.6	8.4	16.9
Contract enforcement			
Procedures required	46	35	35
Time (months)	47.3	13.5	7.7
Property registration			
Procedures required	6	4	7
Time (months)	2.1	1	0.4
Closing a business			
Recovery rate (cents on the dollar)	11.6	35.8	81.2
Time (months)	120	20.4	18

SOURCE: World Bank (undated, 2010a).

The World Bank survey is based on laws and regulations, and the cost indicators in the table are ostensibly backed by official fee schedules. Respondents to the World Bank survey filled out written surveys and provided references to the relevant laws, regulations, and fee schedules, aiding in data checking and quality assurance. The authors of the present monograph have not had an opportunity to validate the World Bank results.

gests that it remains the least urbanized of the emerging Asian economies. According to analysts from the McKinsey Global Institute, only 29 percent of the Indian population currently lives in cities, compared with 40 percent in China and 48 percent in Indonesia. The McKinsey analysts project that the level of urbanization will increase to only 37 percent by 2025 in India. Finally, they note that while more Indians are completing secondary and higher education, India's educational system remains severely strained and that opportunities for schooling vary widely, as does the quality of schooling. Indeed, nearly all of the business group authors stress educational inequality in India as a significant impediment and a relative disadvantage in comparison with the educational condition in China.[11]

Studies by International Organizations

We included seven studies by international organizations in the set of 27 studies.[12] Table 3.4 summarizes the international organizations' GDP growth estimates for China

Table 3.4
China-India GDP Growth Estimates by International Organizations, 2020–2025 (%/year)

	China	India
Low	5.9	5.2
High	9.0	8.0
Average	6.8	6.2
Variance	1.1	0.9
N (observations)	7	8
N (studies)	5	6

NOTE: Differences between the numbers of observations and of studies arise because a few of the studies present a range for each high and low estimate rather than a single figure.

[11] Heng Quan, in his paper titled "Income Inequality in China and India: Structural Comparisons," observes Gini coefficients (reflecting socioeconomic inequality) for both China and India since 1980. He shows that China's regional differences were higher than those of India before 1990–1991, reflected in India's lower Gini coefficient. But India's coefficient has increased since 1991, evidently exceeding that of China.

[12] The seven studies in the international organization cluster are Cooper (2005), Gupta (2002), Rodrik and Subramanian (2004), Shiyang (2007), U.S. Department of Energy, Energy Information Administration (2008), Winters and Yusuf (2007), and World Economic Forum (2006).

and India in the 2020–2025 period. This cluster of studies projects higher growth estimates than those made by the business cluster, but similar estimates to those in the academic cluster. Whereas the business cluster's range of growth estimates is wider for India than for China, the international organization cluster shows a wider range in the estimates for China than for India. The variance of the estimates for China is larger for the international organization cluster than for the business cluster, although still lower than for the academic cluster.

An anthology published by the World Bank (Winters and Yusef, 2007) provides several analytic models for assessing developments in the Chinese and Indian economies and their impact on global markets through 2020. In addition to providing forecasts of the two countries' economic growth, this book analyzes what would occur if China were to grow at an annual average rate of 6.6 percent, and India at 5.5 percent, through 2020. Several essays explore other facets of China and India's growth, including effects on the geographical location of global industry, changes in the international financial system, effects on the global environment, and the relationship between growth and governments.

A paper from the International Monetary Fund (Rodrik and Subramanian, 2004) estimates India's annual GDP growth during the 2020–2025 period using a growth-accounting model based on inputs of capital and labor and increases in factor productivity. Rodrik and Subramanian acknowledge that their estimates may be low if India succeeds in expanding and improving its educational system. They note that India's productivity growth has benefited from its stock of highly educated people, although they do not provide much supporting evidence. They also acknowledge that their growth forecasts rely on continuation of effective economic and social reforms in India. Rodrik and Subramanian also contend that, unlike China, India already has strong economic and political institutions, so that further reform need not be burdensome. Instead, they suggest that India "has done the really hard work of building good economic and political institutions—a stable democratic polity, reasonable rule of law, and protection of property rights," concluding that "countries with good institutions do not in general experience large declines in growth."

As previously noted, not all of the authors and clusters of studies agree with this judgment. Some of the other studies contend, instead, that the effectiveness of India's institutions leaves much to be desired (e.g., Poddar and Yi, 2007).

Similarities and Differences Among the Clusters

As indicated in Table 3.1 and the preceding sections, forecasts of the absolute and relative macroeconomic performance of India and China in the 2020–2025 period betoken deep uncertainty. Estimates of both China and India's annual economic growth over this period vary by a factor greater than two (3.8–9 percent for China and

2.8–8.4 percent for India) across the 27 studies in our meta-analysis. The range in forecasts for the United States and most OECD countries would be less than half as large.

Table 3.5 and Figure 3.2 summarize how the three clusters of studies differ from each other. For example, the widest variation in growth estimates for both China and India comes from the academic cluster. Table 3.5 also shows that the estimates of growth rates emanating from scholars at international organizations tend to be the highest among the three clusters, while the estimates from the business cluster show the widest difference in the growth estimates between India and China. Furthermore, the business cluster projected distinctively higher growth estimates for India than for China.

Table 3.5
China-India GDP Growth Rate Estimates by the Three Clusters, 2020–2025 (%/year)

	Country	Minimum	Maximum	Average	Variance
Academic institutions	China	3.8	9.0	5.5	2.6
	India	2.8	7.2	4.3	2.4
Business organizations	China	4.5	5.2	4.7	0.1
	India	5.4	8.4	6.3	1.1
International organizations	China	5.9	9.0	6.8	1.1
	India	5.2	8.0	6.2	0.9

Figure 3.2
China-India GDP Growth Rate Estimates by the Three Clusters, 2020–2025

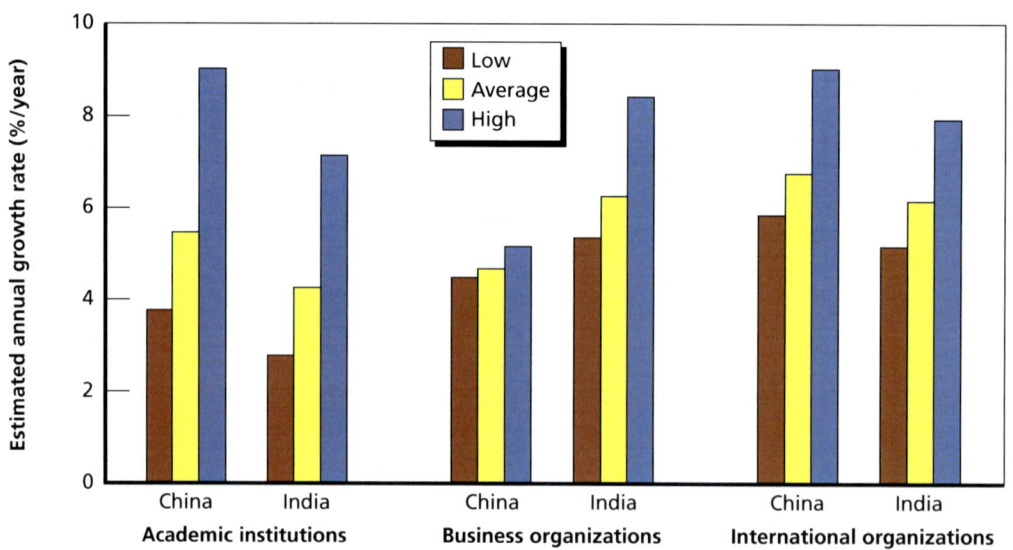

The contrasting forecasts of the three clusters are portrayed visually in Figure 3.3. In the figure, India's annual growth is indicated along the y-axis, and China's on the x-axis. The three rectangles show the distribution of the summary statistics for, respectively, the business cluster (in orange), the academic cluster (in blue), and the international cluster (in lavender). The means for each cluster are indicated by the correspondingly colored dots in each rectangle. The lowest growth estimates for India and China for each cluster appear at the lower left (or southwest corner) of each rectangle, and the highest growth estimates appear at the upper-right (northeast) corner of each rectangle. The x's shown in Figure 3.3 represent, from top right to lower left, the high, average, and low China-India growth estimates for the pooled set of all 27 studies included in the meta-analysis.[13]

Interpreting, let alone explaining, the notable differences among the three clusters is bound to be conjectural. For example, the widest variances characterizing the academic cluster's estimates might plausibly be attributed to greater awareness by the scholarly community of the enormous sources of uncertainty affecting economic forecasts a decade-and-a-half into the future. Perhaps another influence contributing to

Figure 3.3
Summary of the Average, High, and Low Estimates, for All 27 Studies and by Cluster

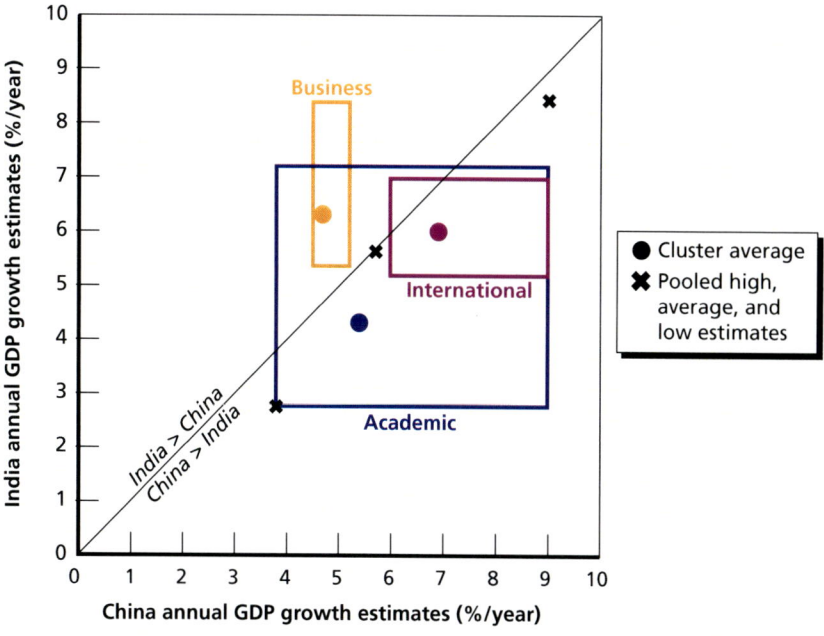

---

[13] We are indebted to RAND colleague Michael Mattock for this graphic. Mr. Mattock was one of the two reviewers of this monograph.

this spread may lie in differences in worldviews—that is, a division among academic economists between those who are professionally inclined to favor central planning and an expanded role of government in economic development, and those who are inclined to favor greater reliance on free markets, market-based pricing, and decentralized innovation in determining resource allocation. Those on the former side may tend to see a rosier outlook for China, while those on the latter side may be inclined to view India's prospects more favorably. The result of these differing dispositions and behaviors may be the widened variance of their respective forecasts.

By similar reasoning, it may be presumed that studies of economic growth sponsored by business organizations might tend to be led by economists inclined toward market-based development. Hence, such studies are probably more likely to view India's democracy, rule of law, and legally protected property rights as constituting a more propitious environment for business innovation and long-term economic growth than that provided by China's one-party autocracy. Consequently, it is not surprising that the highest growth estimates for India relative to China come from the business-sponsored studies included in the meta-analysis.

Finally, that the forecasts made by the international organizations' studies show a marked advantage in China's expected growth relative to India's may plausibly be attributed to China's more prominent role in international trade and investment markets relative to India. As a consequence, one might expect international organizations to be particularly cognizant of and sensitive to this in their estimates of the two economies' growth over the next 15 years, resulting in the relatively buoyant forecasts for China.

Underlying and contributing to the wide differences in forecasts are significant differences in the assumptions made by the forecasters. For example, some of the forecasts simply assume a continuation of recent growth trends in both countries, extrapolating linearly to forecast the 2020–2025 period. Other forecasts focus especially on demographic trends and especially trends in labor supplies that inhere in the current circumstances of the two countries' population cohorts and fertility rates. Still other forecasts build their estimating models on assumptions relating to energy prices and the heavy dependence of the two countries on fossil fuel imports. Further, some of the forecasts make assumptions about the prevalence of macroeconomic stability, economic openness, the quality or inequality of educational opportunities, and the integrity of economic and social institutions. Embedded in most of the studies that use the neoclassical model described earlier are simplifying and arguable assumptions about constant returns to scale and competitive markets.

In turn, these assumptions and the selectivity of their focus affect the inputs to the analytic models that the authors use in generating their respective forecasts. In the process, the forecasts ignore cyclical fluctuations around long-term trend estimates. They ignore the possibility of such major adverse shocks as political disturbances, natural disasters, or military conflict and, on the optimistic side of the spectrum, the possibility of a major technological jump that might trigger a new wave of innovation in either China or India.

Always, implicit, and sometimes explicit, in the respective forecasts is a recognition by the authors that China and India have taken quite different paths in pursuit of their economic development. China has emphasized the expansion of labor-intensive manufacturing, while India has charted a path from agriculture to high-end services with a limited increase in the manufacturing sector. In sum, the wide range of the estimates reflects both the assumptions and behavioral dispositions of the forecasters, the issues they focus on as well as those they ignore, and the deep uncertainties that surround forecasts over the next decade-and-a-half.

Five Growth Scenarios and Concluding Observations

The meta-analysis discussed in the preceding sections displays quantitatively the profound uncertainties that pervade attempts to forecast how two such dynamic and complex systems as the economies of India and China will fare over the next 15 years. This uncertainty pervades the 27 studies encompassed in our analysis, whether they are examined in the aggregate or within the three separate clusters of the academic, business, and international organization studies.

In this section, we contrast five scenarios consisting of different pairings of the forecasts for the two countries: a scenario in which both countries grow at their respective average estimates, and scenarios that show the four combinations of the separate high and low growth estimates for China and India. On the implicit but plausible premise that many of the major factors affecting the economic performance of China and India (e.g., their respective fiscal and monetary policies, education policies, business regulatory policies, etc.) are uncorrelated with one another, these starkly contrasting high-low scenarios can serve two purposes: first, to highlight (and in some sense magnify) the uncertainties that emerge from the meta-analysis, and second, to provide a basis for contingency planning for policymakers. More specifically, the challenges that U.S. policymakers face will be very different depending on which of the contrasting scenarios ensues. That said, it should also be noted that the most appropriate policy responses to the contrasting scenarios are more likely to involve adjusting to them, rather than shaping them. We do not mean to suggest, for example, that U.S. policy is without some influence on which scenario occurs, but rather that the extent of such influence, as well as of the resources that the United States is likely to be willing to deploy to affect scenario outcomes, affects the scenarios only at their margins rather than their defining cores.

Figure 3.4 shows the five contrasting GDP growth pairings between China and India in 2020–2025 under the five contrasting scenarios.

Figure 3.4
Five GDP Growth Scenarios, India and China, 2020–2025

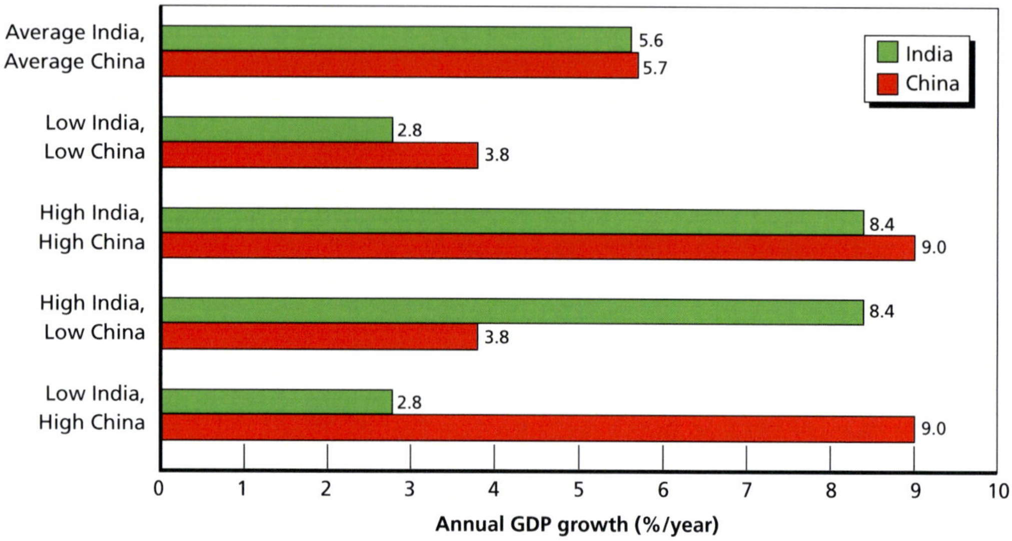

Figures 3.5 and 3.6 show the GDPs for India and China in 2025 in terms of market exchange rates (Figure 3.5) and PPP conversion rates (Figure 3.6).

As Figures 3.5 and 3.6 indicate, only in the scenario in which high growth in India is paired with low China growth does India's GDP approach China's. In the four other scenarios, China's predominance is decisive. This outcome is the same whether conversions are calculated with market exchange rates or PPP rates.

Turning to a more qualitative aspect of the China-India assessment, Table 3.6 distills from the meta-analysis our judgment about the advantages and disadvantages of China and India in their respective institutional and other circumstances.

Whether and to what extent the factors listed in the preceding table will enable India to move closer to, or ahead of, China after 2025 is worthy of separate consideration.

Figure 3.5
Five Scenarios: GDPs of China and India in 2025, Market Exchange Rates

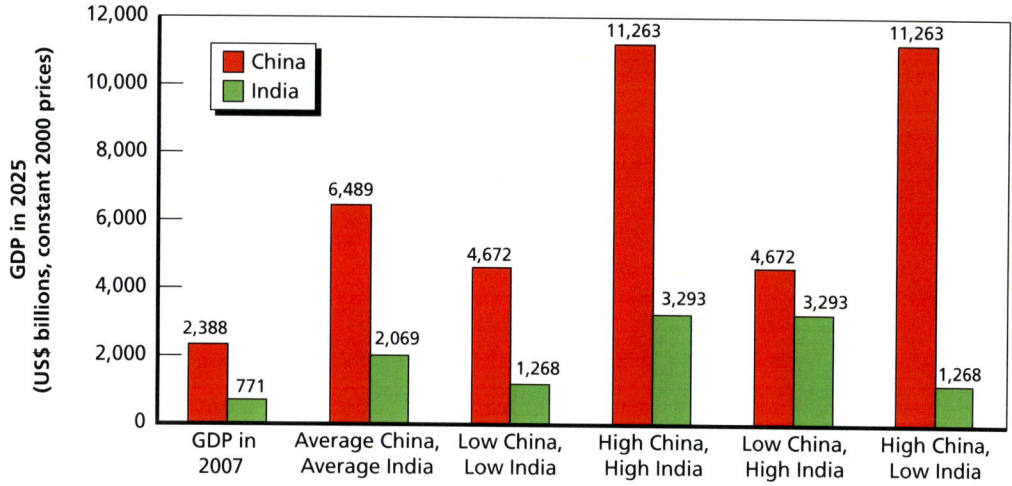

NOTE: Conversion to market exchange rates based on the World Bank's World Development Indicators (World Bank, no date).

RAND *MG1009-3.5*

Table 3.6
Some Qualitative Factors Affecting China and India's Performance

Factor	Advantage of China or India?
Democracy/rule of law	India
Information technology and service skills	India
Institutions	India
Property rights	India
Productivity growth	China
Foreign investment in and by each country	China
Infrastructure	China

Figure 3.6
Five Scenarios: GDPs of China and India in 2025, Purchasing Power Parity Conversion Rates

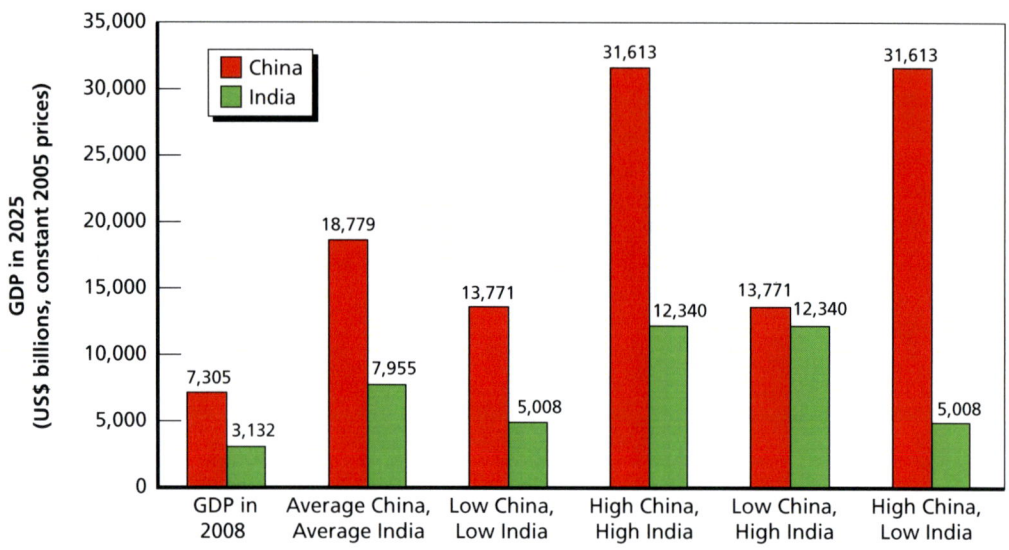

CHAPTER FOUR

Science and Technology

Introduction

Scientific research, invention, and innovation are key drivers of economic growth. During the past three decades, R&D investment in the world economy has been concentrated in 30 countries, all of which are members of the OECD (National Science Foundation, 2007a). Nevertheless, global R&D structures and innovation are undergoing transition, and there is strong evidence that patterns of research and innovation are changing (OECD, 2008a).

The main dimensions of this change are (1) the absolute growth in total R&D activity, (2) the rise of particular emerging market economies—especially China and India—in terms of not only their economic size, but also the intensity of their science and technology efforts, and (3) use of a growing diversity of policy instruments to foster innovation (OECD, 2007a).

Non-OECD countries account for a growing share of global R&D investment: 18.4 percent in 2005, compared with 11.7 percent in 1996 (OECD, 2008a). Meanwhile, the U.S. and EU shares in global R&D have decreased by 3 percent and 2 percent, respectively. Other R&D indicators, such as publications, also reflect a similar reallocation of R&D effort. OECD countries have retained their dominance in science and engineering (S&E), producing 85 percent of S&E publications from 1993 to 2002—but two-thirds of the remainder were generated by China, India, Brazil, Russia, and South Africa. The growth in S&E publications from these five non-OECD countries has been much more rapid than that of the rest (NSF, 2007a).

Innovation and investment depend on more than simply spending more money. They also benefit from development of a sound legal environment to protect intellectual property, investment in higher education to increase the supply of highly skilled human resources, and a favorable and predictable regulatory environment that contributes to the incentives for as well as protection of intellectual property.

The purpose of this chapter is to provide a brief overview of S&T infrastructure and performance indicators for China and India, as well as forecasts for these indicators through 2025. Our forecast methodology is based on an "accounting model" that uses estimates of each country's GDP and its growth as exogenous variables; these estimates are derived from the meta-analysis in Chapter Three. The model is linear, with constant returns to scale and with slopes estimated from current data. Thus, the model can be used to generate simple and clean quantitative estimates, although serious questions can be raised about whether the model's assumptions exclude important aspects of the dynamic processes that are involved by tying our estimates too closely to GDP growth. Our forecasts are presented in U.S. dollars converted from rupees and renminbi using PPP conversion rates. The outlays in local currencies could also be converted using market exchange rates, as was done in Figure 3.5 in Chapter Three.

In the next section, we highlight several global trends in S&T and how they are influencing China and India. The global trend indicators that we use include R&D spending, human resources, and patents, especially triadic patents (those that are filed in the U.S., EU, and Japanese markets).

Highlights and Global Trends

In this section, we highlight several major trends affecting R&D globally in general and in China and India in particular. Technology policies are difficult to measure. Thus, analysis and comparisons have to rely on a set of indicators to approximate different phases of these policies. There are two types of indicators: input and output.

Input indicators include financial and human resources in R&D activities. Investment in R&D—usually called R&D intensity—is approximated by gross expenditures in R&D (GERD) as a percentage of GDP. GERD is usually decomposed into four performing sectors: business (BERD), government (GOVERD), higher education (HERD), and private nonprofit (PNPERD). Human resources include researchers and personnel involved in research activities. To ensure comparability, human resources are measured on a full-time equivalent scale. Researchers include people with a Ph.D. in S&E.

Clearly, the contributions of R&D to the society are the ultimate measure of R&D's output. As proxy indicators of R&D output, we use the number of patents and publications produced by each country's researchers.

Global Trends

Studies of the sources of macroeconomic growth highlight the relevance of information technology (IT) and its contribution to multifactor productivity (OECD, 2008a, p. 20). OECD (2008a) reports that the widening disparities in growth trends over the past few decades are largely due to the size of the information and communication technology (ICT) industries and the pace of IT adoption by other industries. Countries' ability to produce and incorporate innovation will determine their future economic growth.

Financial Indicators. Gross domestic expenditure in R&D (GERD) as a percentage of GDP—also known as research intensity—has been increasing in real terms at a 3.2 percent average rate worldwide for the past three decades (OECD, 2008a). But, as a consequence of the financial crisis, GERD is forecasted to slow down. For instance, in the United States the rate of growth slowed to 1.3 percent in 2008.

Table 4.1 shows GERD and its four components as a percentage of GDP for select countries for the 1995–2005 period. Figure 4.1 highlights the preponderant share of India's R&D spending represented by government.

Business R&D (BERD) is the major component of GERD in industrialized countries (OECD, 2008a, p. 11). Although it was the main driver of GERD in the 1990s, BERD has diminished since the beginning of the 21st century, with the high-tech recession in early 2000 and thereafter. BERD is higher in Japan, South Korea, and the United States than it is in the European Union (EU). U.S. BERD is nearly 2 percent of GDP, while in the EU it is only half of this intensity, well short of the Lisbon Strategy's BERD target of 2 percent of GDP.[1]

BERD is related to the composition of industry, because some business sectors are more R&D-intensive than others. BERD is also related to the presence and relative

Table 4.1
Composition of Gross Domestic Expenditure in R&D, by Country (average from 1995 to 2005 as a percentage of GDP)

Country	GERD	HERD	BERD	GOVERD	PNPERD
Japan	3.1	0.4	2.2	0.3	0.1
South Korea	2.5	0.3	1.9	0.4	0.0
United States	2.6	0.3	1.9	0.3	0.1
EU15[a]	1.8	0.4	1.2	0.3	0.0
India	0.8	0.0	0.2	0.6	0.0
China	1.0	0.1	0.6	0.3	0.0
Taiwan	2.1	0.3	1.4	0.5	0.0

SOURCE: Based on OECD (no date, 2008b, 2008c).
[a] EU15 refers to Austria, Belgium, Denmark, Finland, France, Germany, Greece, Ireland, Italy, Luxembourg, the Netherlands, Portugal, Spain, Sweden, and the United Kingdom. These were the member countries in the European Union prior to the accession of ten candidate countries in May 2004.

---

[1] The "Lisbon Strategy" refers to the strategy developed by the European Council in its meeting in Lisbon in March 2000. The strategy, aimed at making the European Union the most competitive region in the world, has three pillars: economic, social, and environmental. The economic pillar emphasizes the urgency to adapt "constantly to changes in the information society" as channels to remain competitive and to achieve full employment by 2010.

Figure 4.1
Component Shares of R&D Expenditures for Selected Countries, 1995–2005

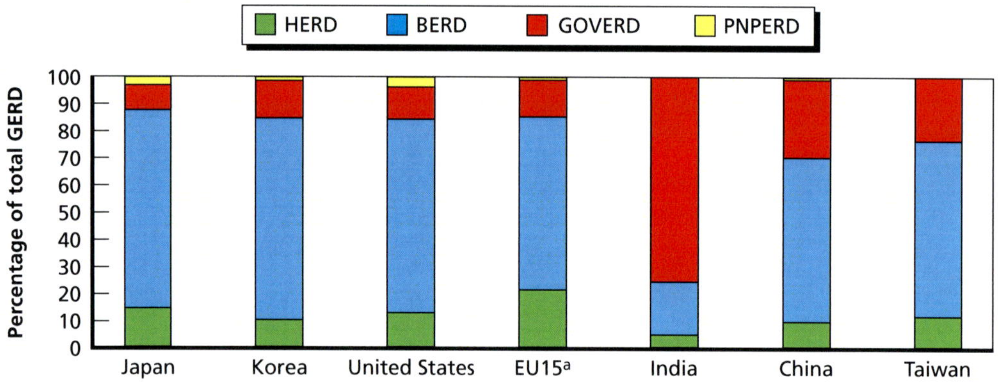

SOURCE: OECD (no date, 2008b, 2008c).
NOTE: Percentages indicate the average over the 1995–2005 period.
[a] EU15 refers to Austria, Belgium, Denmark, Finland, France, Germany, Greece, Ireland, Italy, Luxembourg, the Netherlands, Portugal, Spain, Sweden, and the United Kingdom. These were the member countries in the European Union prior to the accession of ten candidate countries in May 2004.

RAND *MG1009-4.1*

shares of large firms, which tend to invest more in R&D than do small ones. Business R&D intensity has grown in most Asian countries. China's BERD is currently half as large as that of the EU due to China's increase in business investments in the past three decades.

Public investment in R&D has two components: intramural spending and higher education funding. Most EU countries have increased the government-financed component (GOVERD) of R&D activities, with the objective of increasing GERD to 3 percent of GDP by 2010, consistent with the Lisbon Strategy. GOVERD averages 0.65 percent in the OECD area, with an overall decreasing trend that reflects, in part, a shift from direct to indirect support of R&D in the business sector (OECD, 2008a).

HERD intensity varies in accord with the fields in which higher education spending is concentrated. For example, 85 percent of Taiwan's HERD spending is on engineering and mathematics, whereas Taiwan's spending on social sciences is almost negligible (IECD, 2008a).

When GERD is broken down by domestic versus international funding, international sources clearly emerge as a sizable component. This is due to the increasing internationalization of R&D activities. Overall, R&D funding from abroad (through private businesses, public institutions, or international organizations) has increased steadily in the past 15 years. External funding as a percentage of GERD ranges from 11 percent to 26 percent and has increased sharply since the mid-1990s. This internationalization mirrors the increased collaboration between countries, which has in turn led to increases in co-invention on patents and co-authorship of scientific publications (OECD, 2008a, p. 30).

Human Resources. The availability of skilled human resources in S&T is a measure of a country's ability to innovate, sustain, and create high-technology jobs. Highly skilled people create and diffuse innovation and are as important as capital inputs. Overall, human resources for science and technology (HRST) has increased more rapidly than total employment in the past few decades (Figure 4.2). This trend fits well with the increasing demand for highly qualified workers for all sectors of the economies and not just for research-related activities in particular. India's decrease in HRST during the past decade is a puzzling anomaly that may be due to changes in the way HRST is measured in India.

The S&E degrees of people employed in engineering, manufacturing and construction, life sciences, physical sciences, agriculture, mathematics, and computing as a proportion of the total degrees are a loose indicator of the potential contribution to technology innovations in S&T. As Figure 4.3 shows, this percentage varies across countries, with Taiwan heavily concentrated in S&E diplomas (more than 80 percent of total doctorates), and the EU being the region with the lowest proportion in the sample. The remainder of the sample has around 40 to 50 percent of their graduates in the innovation-related fields. The low percentage of S&E graduates is a concern for

Figure 4.2
Growth of HRST Occupations and Total Employment, 1995–2006, by Country

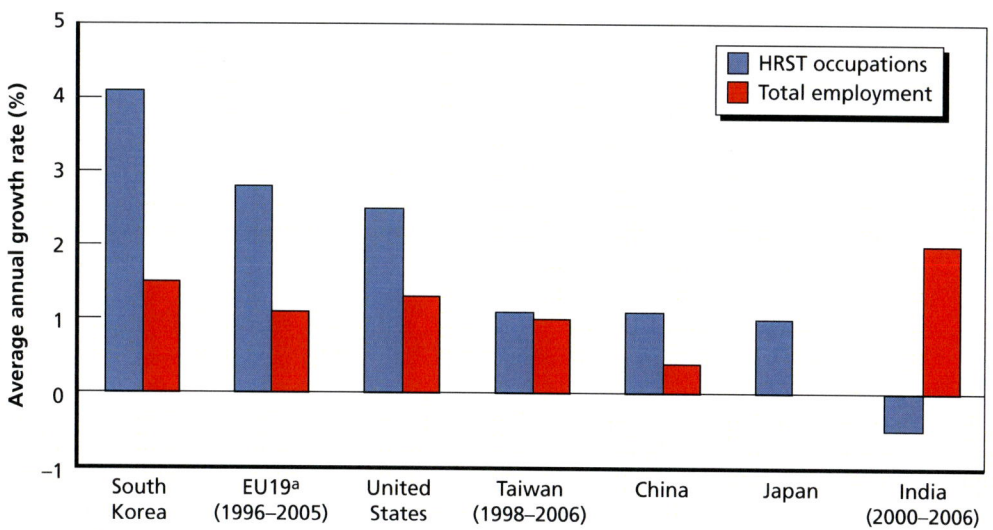

SOURCE: OECD (no date); OECD (2008a), Figure 32; OECD (2007c).
a EU19 refers to Austria, Belgium, the Czech Republic, Denmark, Finland, France, Germany, Greece, Hungary, Ireland, Italy, Luxembourg, the Netherlands, Poland, Portugal, the Slovak Republic, Spain, Sweden, and the United Kingdom.
RAND *MG1009-4.2*

Figure 4.3
Doctoral Degrees in S&E as a Percentage of Total Doctoral Degrees, by Country and Year

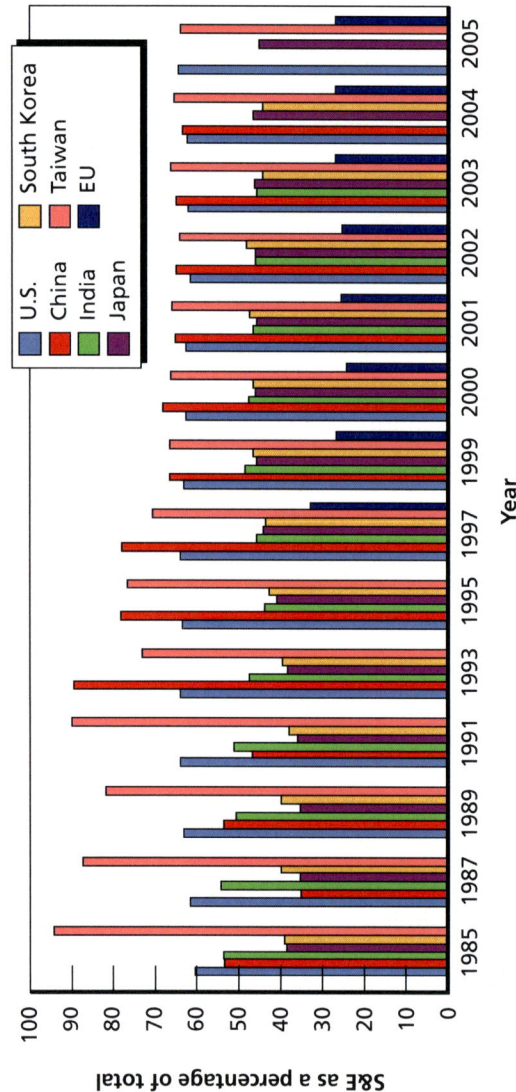

SOURCE: National Science Board (2006), Appendix Tables 2-43 and 2-38.

NOTES: Data for doctoral degrees use the International Standard Classification of Education (ISCED 97), level 6. S&E data do not include health fields. Japanese data include thesis doctorates, called *ronbun hakase*, earned by employees in industry. Japan includes computer sciences in engineering.

RAND *MG1009-4.3*

many countries, according to OECD (2008a), and many of them have implemented policies to increase human resources in science and technology (OECD, 2008a, p. 98).

Scientific Production. R&D output is generally reflected in two types of indicators: patents (applied research) and published journal articles (basic R&D). Countries that spend more on R&D (such as the United States, Japan, Germany, and France) have higher propensity to patent (OECD, 2008a; Bernardes et al., 2006).

Publications. The numbers of articles published provide information about the size and the focus of research activity. Worldwide number of publications has been increasing especially in the countries that have low initial levels. Figure 4.4 shows that while the United States has remained stable in number of publications since 1995, most Asian countries have dramatically increased the total number of articles published and counted by Thomson Reuters Web of Knowledge (formerly the Institute for Scientific Information [ISI]), from 51,000 in 1988 to 135,000 in 2003 (NSF, 2007b, p. 21). China has quadrupled its S&E publications in a decade in 2003 and, as a result, China's share in Asia has grown from 11 percent to 22 percent, the same level as Japan's. India is recovering since 2001 after a long stagnation period (NSF, 2007a, p. 1).

With respect to national strategies, there is no single or unique "mode" of innovation. Even if common innovation patterns have been identified, there are major national differences in patterns of competitive and comparative advantage. Innova-

Figure 4.4
Growth in S&E Publications, 1995–2005, by Country (index 1995 = 100)

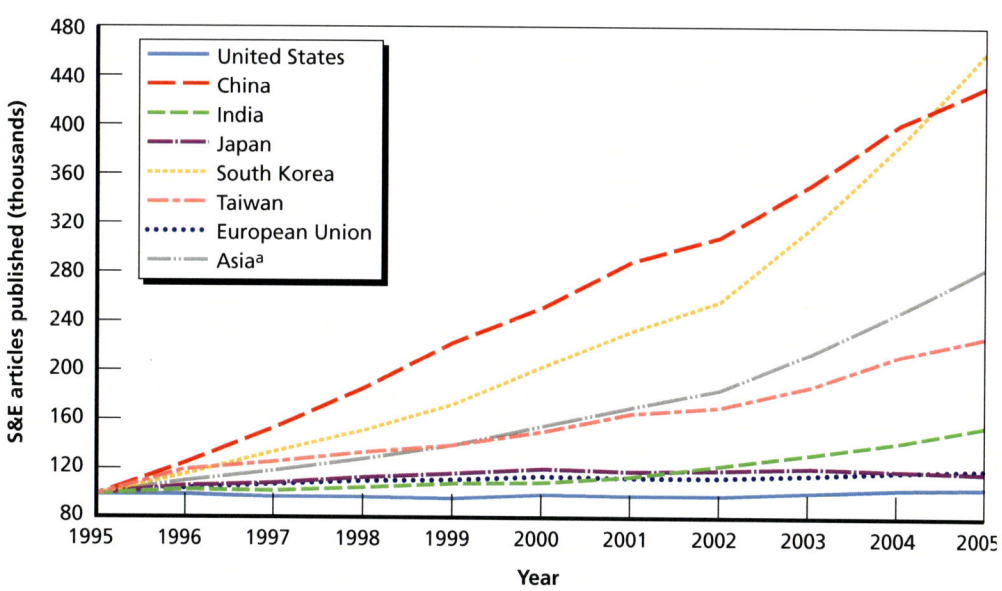

SOURCE: National Science Foundation (2008, Appendix Table 5-34).

[a] Asia includes China, India, Japan, South Korea, and Taiwan.

RAND *MG1009-4.4*

tions in firms include not only technological innovation and generation of technology but also nontechnological innovations, such as advanced management techniques (for example, total quality management). Policies to foster innovation should extend to all sources of innovation that may boost productivity growth.

Patents. Patents indicate a country's innovative capacity, but they are an imperfect measure because of the differences in national patent laws and the fact that some innovations are not patentable. For this reason, the analysis of patenting tends to focus on patent filings in the United States, since the United States tends to attract innovations from around the world. Half of the patents issued in the United States were filed for foreign residents. Nevertheless, this is still an imperfect measure, in part because countries may differ systematically in their propensities to apply for U.S. patents depending on the extent of their commercial ties.[2] In the analysis presented here, we examine triadic patents: patents that are filed at the European Patent Office (EPO), the USPTO, and the Japan Patent Office (JPO) for the same invention, by the same applicant or inventor.

Filing for patents in three different markets (United States, EU, and Japan) is costly in terms of paperwork, time, and money but gives access to the most lucrative markets. For this reason, a triadic patent is a good proxy for the patent's economic value. Figure 4.5 demonstrates that Asian economies increased their share in triadic patent families; South Korea, Japan and China were the countries that led the growth. Asian countries' global share of triadic patents increased between 1995 and 2005. During this period, South Korea's share increased by 5 percentage points, Japan's increased by 2, and China's increased by 0.7 (OECD 2008a, p. 43). Although India has had some increase since the 1990s, its contribution to the world total is still modest (OECD, 2008a, p. 43).

China

China is the third-largest R&D spender worldwide after the United States and Japan, with an annual GERD growth rate of 18 percent since 2000 (OECD, 2007b, p. 14).

China's BERD has increased from 0.25 percent of GDP in 1996 to around 1.01 percent in 2006. This increase reflects, in part, the restructuring of formerly state-owned enterprises and is consistent with China's decrease in GOVERD. China's annual HERD growth rate (9–10 percent) is ranked second in the world since 2001 (OECD, 2008a, p. 29).

---

[2] In addition, patents are indicators of research investments made years earlier (it generally takes between 18 months and 5 years to obtain a patent). Furthermore, statistics differ according to where patents are filed. For instance, while analysis of U.S. Patent and Trademark Office (USPTO) filings indicates a scarce-to-nil contribution from Chinese and India researchers, statistics from the Swiss-based World International Property Organization show a fourfold rise in patent filings by Chinese and Indian inventors between 1995 and 2006. These statistics indicate that 5.5 percent and 8.4 percent of patents listed one or more investor in India and China, respectively.

Figure 4.5
Triadic Patents as a Percentage of World Total, 1985, 1995, and 2003, by Country

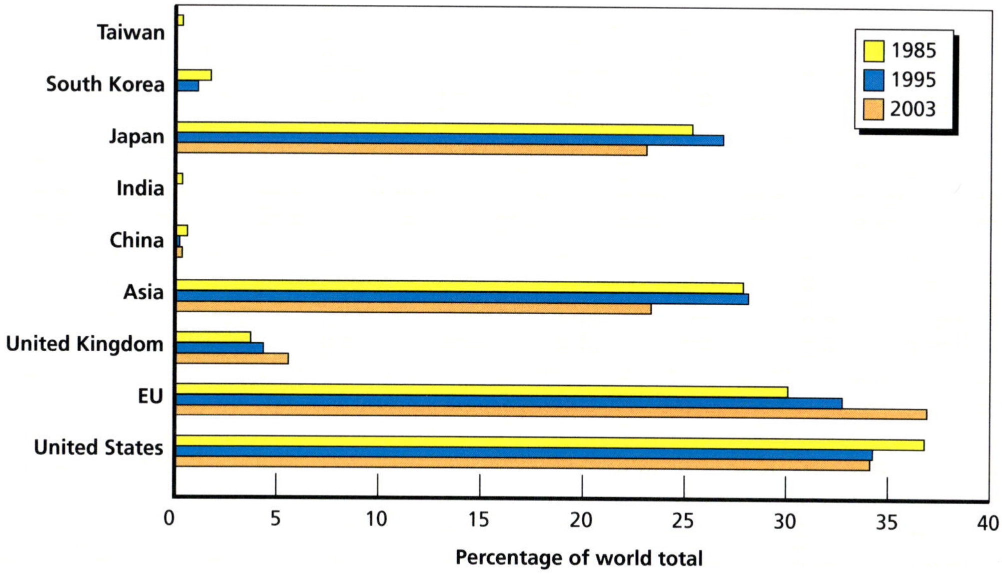

SOURCE: National Science Board (2008).
RAND *MG1009-4.5*

In its 11th Plan, China has set the following goals, to be accomplished by 2020:

- raise GERD to at least 2.5 percent of GDP
- raise total factor productivity (TFP) to 60 percent of GDP growth
- number of Chinese invention patents to rank in the world's top five
- number of internationally cited Chinese articles to rank in the world's top five.

India

India's GERD is around 0.8 percent of GNP, compared with more than 2 percent in the developed countries (Ministry of Science and Technology, Department of Science and Technology, Government of India, no date).

Breaking down the components of India's GERD, 70.5 percent is GOVERD (62 percent central and 8.5 percent state government), 4.2 percent is HERD, and 25.3 percent is BERD (5.0 percent from public-sector industries and 20.3 percent from private-sector industries) (Ministry of Science and Technology, Department of Science and Technology, Government of India, no date). India's 11th five-year plan, covering 2008 to 2013, calls for India to triple its investment in S&T to 2.5 percent of GDP (Table 4.2).

Table 4.2
India's 11th Plan (2008–2013)

	Five-Year Plan		
	1997–2002 (actual % values)	2002–2007 (actual % values)	2008–2013 (% goals under India's 11th Plan)
Macroeconomic growth	5.5	7.2	9
Industry	4	8.3	10.5
S&T as % GDP	0.7	0.8	2.5

Quality and Comparability of S&E Diplomas and the Skilled Diaspora

China graduates 600,000 engineers per year, and India 350,000;[3] by comparison, the United States graduates roughly 70,000 undergraduate engineers annually.[4] In November 2005, the United States called for increasing this figure to 100,000.[5] Evidence from executives doing business in India and China indicates that both countries still have shortages despite the number of graduates (McKinsey Global Institute, 2005). Part of the problem is the quality of the statistics. While American graduation statistics are accurate, Chinese and Indian statistics are quite fragmented and imperfectly comparable.

Quality of Engineers Compared

The meaning of the word *engineer* varies by country. In China, the word does not translate well into different dialects and has no standard definition (Wadhwa et al., 2007): "A motor mechanic or a technician could be considered an engineer, for example." Wadhwa et al. (2007) provide several examples of how Chinese statistics count any degree related to information technology as an engineering degree, including "any bachelor's degree with 'engineering' in its title," and regardless of the number of years needed to obtain the degree.

The increase in engineering graduation rates in China was originated by the Chinese government's initiatives, starting in 1999, to transform science and engineering education into mass education. Higher enrollments in schools, reductions in teacher salaries, and reductions in the number of technical schools and teachers led to dramatic increases in class sizes and deterioration in the quality of education provided by

---

[3]   Chinese information comes from the Ministry of Education (MoE) and from the China Education and Research Network (CERN). India's data are from the National Association of Software and Service Companies (NASSCOM) and the All India Council for Technical Education (AICTE).

[4]   According to the Department of Education's National Center for Education Statistics.

[5]   This was part of the Democrats' Innovation Agenda presented at the House of Representatives in 2005.

Chinese schools.[6] Reportedly, multinational firms operating in China feel comfortable about hiring graduates from only 10 to 15 of its universities. According to China's National Development and Reform Commission, 60 percent of university graduates in 2006 were not able to find work. As a consequence, the Chinese Ministry of Education announced in June 2006 that it would begin to reduce enrollments.[7] This announcement was part of more comprehensive reforms in tertiary education intended to increase productivity in teaching and research; academic personnel will also be recruited through open advertisements and on a merit basis, while a system of performance management will be introduced in higher education institutions (Gallagher et al., 2009, p. 6)

India offers a very different scenario. Engineering education in India has been market-driven and characterized by few regulatory bodies and an inefficient and heavily politicized public education system. Ongoing debates in the country focus on the demand for caste-based quotas for more than half of the available seats in public institutions. Due to the inefficiencies in the public sector, national innovation is based on the private educational institutions. In 2004, India had 974 private engineering colleges and only 291 public institutions. Moreover, private training centers across the country have flourished and serve corporations that need to train employees (and job seekers). Among the public universities, quality varies widely; some centers, such as the Indian Institutes of Technology, provide high-quality education, but they graduate only a small percentage of India's engineers.

Despite the heterogeneity in quality, multinational executives feel more comfortable about hiring Indian engineering graduates. A 2005 survey from the McKinsey Global Institute (2005) of 83 globally based multinationals reports that 80.7 percent of U.S. engineers were employable, whereas only 10 percent of Chinese engineers and 25 percent of Indian engineers were similarly employable (Gereffi et al., 2008, p. 21). The main reasons for the low employability figures for China's and India's engineers were that the engineers' background included little or no hands-on experience despite solid theoretical training, poor English-language ability, and overall communication and cultural style (McKinsey Global Institute, 2005). These employability numbers have direct bearing on competitiveness in global labor markets. We refrain from making an adjustment of this sort in our model because we assume that quality will improve in the future. However, later in this chapter we present an exercise that allows for such a "quality" adjustment.

---

[6] Only Tsinghua and Fudan, elite universities, were allowed to decrease class sizes when they reported quality problems (Wadhwa et al., 2007).

[7] China's 11th Plan contemplates achieving universal nine-year compulsory education, as well as developing preschool and special education.

Graduate and Postgraduate Engineering

Based on on-site interviews, Wadhwa et al. (2007) report that the business executives preferred to hire graduates with master's or Ph.D. degrees, though they did not mandate Ph.D. degrees for research positions. Getting either a master's or Ph.D. degree is easier for Chinese executives than for Indian executives. For the United States, on the other hand, the problem is the shortage of native students completing master's and Ph.D. degrees. This could become even more systemic, especially as both countries' economic growth improves and they become more competitive in terms of salary and opportunities for career development.[8,9]

Emerging Models of Diaspora Mobilization

There has been a dramatic restructuring of how corporate R&D is performed (from in-house to elaborate outsourcing) and where R&D resources are spent. A key trend is that newly industrialized countries (India, China) are emerging as a preferred location to outsource corporate R&D. Kuznetsov (2006a, 2006b) distinguishes among several types of networks that have resulted from what he calls the broader phenomenon of "diaspora":

1. **Technology and R&D outsourcing networks** (e.g., the Indian "Top Executives" model): Indian executives in major multinationals influence investment decisions to outsource knowledge-intensive operations to India. Pandey et al. (2004) discuss how the "Indian Diaspora" has facilitated growth in knowledge-intensive sectors in the United States, and how this has, in turn, played a decisive role in the emergence of the knowledge-intensive service sector in India.[10]

---

[8]   A recent report (Wadhwa, 2009a), based on interviews of company executives, foreign students, and returnees, hypothesizes that the United States may be experiencing its first brain drain, as highly skilled foreign-born S&E workers leave the United States to pursue opportunities in their home countries. In a 2008 survey of 1,203 Chinese and Indian immigrants who had worked or received their education and then returned to their home country (Wadhwa et al., 2009), the most frequent reason they gave for being willing to return home was to advance their careers (which was also the most frequent reason they gave for coming to the United States in the first place). Many immigrants plan to stay in the United States for only a few years, to acquire education or experience, and expect to have better career opportunities in their home country than in the United States.

[9]   Social, family, and cultural factors reinforce the trend.

[10] Examples of Indian executives making decisions that benefit India as well as the companies they work for include the following:
  - Kanwal Rekhi (Novell) generated contracts for Infosys and other emerging Indian software firms
  - Alok Aggarwal (IBM) managed to install a research center in India under his lead
  - Rajat Gupta (McKinsey) led McKinsey to be a pioneer in subcontracting research services in India and has helped to establish the Indian School of Business Hyderabad
  - Ash Gupta (American Express) was decisive in American Express's decision to open a customer service center in India that employs 5,000 people.

2. **Cross-border investor networks** (e.g., China's "Bamboo Network"): Diaspora members know well the reality of their home country and have access to risk-mitigation strategies. Personal trust between members of cross-border investor networks also reduces transaction costs.

3. **Scanning networks** (Israel, Armenia, India): Diaspora members identify niches and new opportunities, set new strategic directions, and translate global opportunities into business projects.

4. **Brain circulation networks** (China, South Korea): Home countries may provide incentives (such as special technology parks in China) for S&E talent to return.[11]

Science and Technology Forecast: India and China in 2025

Based on (1) the policy objectives described above and (2) GDP growth derived from past and current macroeconomic performance and R&D investment, we have projected the main R&D indicators for China and India. The projections are based on an "accounting model" driven by GDP estimates obtained as an exogenous variable from the meta-analysis summarized in Chapter Three. The model includes parameters with several ratios (e.g., number of researchers = GERD × [GDP/cost per researcher]) that assume that the cost per researcher will remain close to constant in PPP U.S. dollars. The parameters are based on the same information (from OECD and the governments of India and China) as the R&D statistics presented earlier in this chapter. This "constant returns to scale" linear model with ratios estimated from the current data allows quantitative estimates that are relatively simple and clean, though questions can be raised about the underlying assumptions.[12] These are discussed in the next section. Simplicity provides a useful tool for considering where countries may be in terms of prospective demands for HRST and their expected contribution to innovation (as measured by patents and publications).

---

[11] Brain circulation networks have helped in the dramatic transformation of Silicon Valley (Wadhwa, 2009a, 2009b, 2009c). The model consists of a network of highly skilled people with two types of variants: (1) an alliance between a domestic corporation in the emerging country and a multinational firm in the industrialized country or (2) the multinational retaining control over production but carrying it out in the developing country. In the case of India, some initiatives were taken at the government level, with programs such as Ambassador NRI that have come to reinforce these trends. As a result, countries in that position have introduced different models to mobilize their skilled Diaspora and turn "brain drain" into "brain gain."

[12] At early stages of the project, we analyzed the possibility of running some sort of behavioral model for invention. But this type of modeling demands a large amount of micro information for both countries (India and China), which either is not available or does not exist. More specific models that utilize industry-level panel data as well as information about institutional setting to predict level of innovation might provide more accurate predictions than the one proposed here.

To illustrate the effect of changes in the R&D parameters, we provide estimates of alternative scenarios for cost per researcher and productivity per researcher (defined as number of patents per researcher). For both parameters, we assume as the likeliest scenario that both India and China will converge toward levels observed in more developed countries. This scenario is supported by the experience of South Korea, a good benchmark in Asia. Our model and projections could be easily extended to other changes in exogenous parameters, such as differing rates of economic growth.

That technological progress contributes to economic growth is a central insight of the endogenous growth literature. In several of the canonical models, the assumption is that more resources invested in R&D lead to increasing growth rates (Romer, 1990; Aghion and Howitt, 1992). This is in sharp contrast to the traditional neoclassical, exogenous technological progress, growth model. Attempts to establish the direction of causality and, thus, the distinction between exogenous and endogenous growth models, have been the focus of much of the economic growth literature.[13]

Recognizing the conceptual complexities of the issue (e.g., is technological progress exogenous or endogenous?), we estimate the main indicators of S&T of China and India by utilizing a very simple model that considers technological progress as exogenous to economic growth.

Projection

Our goal is to project the size and components of GERD as a proportion of GDP for China and India; the HRST; and the main indicators of output, patents and number of publications. The projection proceeds from the GDP growth rate and then estimates the unknown variables.

The exogenous variable is the rate of GDP growth under alternative scenarios. The unknown variables are as follows:

- GERD in U.S. dollars of 2008 PPP
- GERD components in U.S. dollars of 2008 PPP
- number of full-time researchers (FTRs)
- number of Ph.D. diplomas in S&E
- number of triadic patents
- number of publications in S&E.

The model includes the following parameters and assumptions:

- GERD as a percentage of GDP is based on India's and China's 11th Plans:
  - for India, 2.5 percent in 2013 and 2025
  - for China, 2 percent in 2010 and 2.5 percent in 2025

---

[13] For instance, Jones (1995) argues that in the United States both economic growth and intermediate measures of knowledge generation (such as patents and publications) appear to be constant over time despite large increases in R&D employment.

- GERD components (HERD, BERD, GOVERD, and PNPERD) are determined **by convergence toward the average over the 1995–2005** period for two benchmark countries, Japan and South Korea. We selected Japan as one benchmark because of its early attainment of S&T indicators on par with those of the EU and the United States, and South Korea because it has experienced a rapid increase in its S&T indicators in the last two decades. Although in the long run we expect both China's and India's S&T indicators to converge to those observed in the industrialized world, we feel that over the time frame 2010–2025 it is more reasonable to expect China's and India's S&T parameters to converge to South Korea's current parameters.
- The average cost of an R&D FTR is estimated at 2007 PPP conversion to dollars and remains constant in our cost projections.
- Number of researchers = (total cost of R&D)/(average cost of an R&D FTR).
- Number of Ph.D. diplomas in S&E per number of researchers – Year = The corresponding ratio in 2003.
- Patents per FTR determined for each country **by its average over the 1995–2005** period.

We provide more detailed descriptions of how these parameters are calculated in Appendix B.

Summary Alternative Scenarios. We present four alternative scenarios. Scenario 1 considers the current S&T parameters in China and India. This first scenario may be considered too conservative, due to the expectation that economic growth and integration with the rest of the world will boost changes in behavior in many sectors, including S&T.

For that reason, we consider two alternative scenarios in which we change some R&D parameters to match those of South Korea, one of the countries that we use as benchmarks. In each of these two alternative scenarios, we move only one of the parameters in the model to the benchmark level and estimate the variables of interest according to that change. In Scenario 2, we assume that cost per FTR will converge to the level in South Korea. In Scenario 3, we assume that the numbers of triadic patents and S&E publications per FTR—which we hereafter refer to as "researcher productivity"—for China and India will converge to those of South Korea.

The last scenario, Scenario 4, discounts the researcher productivity numbers based on the sharply different employability rates for engineers from China versus those from India. As discussed previously, McKinsey Global Institute (2005) reports that, in a 2005 survey of multinational executives, only 10 percent of Chinese engineers were considered employable, versus 25 percent for India. Hence, in Scenario 4, we assume an imputed productivity for Indian engineers 60 percent higher than that for Chinese engineers.

The four scenarios are summarized as follows:

- Scenario 1: Current Cost FTR/Current Researcher Productivity
- Scenario 2: South Korea's Cost FTR/Current Research Productivity
- Scenario 3: South Korea's Researcher Productivity FTR/Current Cost per Researcher
- Scenario 4: Number of engineers adjusted according to the employability study by McKinsey Global.

We provide detailed descriptions of the calculations for each scenario in Appendix B. We present the results of our projections in Tables 4.3 and 4.4. Table 4.3 shows the estimated number of researchers, S&E Ph.D.'s, patents, and S&E publications for China and India for each of the four scenarios through 2025; the results in Table 4.3 are also depicted graphically in Figures B.1–B.4 in Appendix B. Table 4.4 shows estimates for GERD and its four components for each country through 2025.

Discussion

The results presented in Tables 4.3 and 4.4 clearly show that, based on the constant-returns-to-scale model, the objectives presented in India's and China's national plans are too ambitious for both countries, especially in terms of the "standard" measures of output, publications and patents. Overall, the estimates seem too ambitious in terms of intermediate outputs—especially human resources, given the current problems both countries face in terms of producing sufficient quantity and/or quality of S&T personnel.

Researchers and other R&D personnel may be working in less-than-ideal conditions in terms of institutional framework and support, because their productivity, as measured by the number of triadic patents and publications per FTR, is still at very low levels. These observed low levels may be rooted in each country's relative isolation from the global economy in terms of innovation and human capital quality and basing its competitiveness on cheap labor. As China's and India's economies become more competitive in terms of S&T graduates, they will improve the environment for innovation as well. The alternative scenarios presented consider the types of changes that we considered reasonable to expect in the 15-year time frame of our analysis.

The projection numbers should be taken with caution, though they are likely to be robust for examining relative comparisons. On the one hand, some assumptions—such as the constant "productivity" of researchers in terms of publication and patents and the invariability of the cost per researcher (especially in China, which exhibits a very low cost)—could be questioned. Nevertheless, the changes China is introducing in its higher education sector aimed at increasing productivity in research and teaching constitute a basis for optimism. On the other hand, this exercise is purely quantitative; it does not incorporate either government's policies, which may affect its portfo-

Table 4.3
S&T Key Indicators Under Alternative Scenarios, China and India, 2008–2025

	India					China				
	2008	2010	2015	2020	2025	2008	2010	2015	2020	2025
Researchers (in thousands)										
Scenario 1	31	84	139	183	240	1,024	1,860	3,070	4,053	5,350
Scenario 2	40	107	177	232	306	383	696	1,149	1,517	2,002
Scenario 3	31	84	139	183	240	1,024	1,860	3,070	4,053	5,350
Scenario 4	31	84	139	183	240	410	744	1,228	1,621	2,140
Ph.D.'s in S&E (in thousands)										
Scenario 1	3	8	14	18	23	17	30	50	66	88
Scenario 2	4	10	17	23	30	6	11	19	25	33
Scenario 3	3	8	14	18	23	17	30	50	66	88
Scenario 4	3	8	14	18	23	7	12	20	27	35
Patents										
Scenario 1	27	72	119	156	205	172	313	516	681	899
Scenario 2	34	92	151	198	261	64	117	193	255	336
Scenario 3	118	318	524	689	906	3,862	7,018	11,580	15,287	20,180
Scenario 4	27	72	119	156	205	69	125	206	272	360
Publications (in thousands)										
Scenario 1	3	9	15	20	26	22	41	67	88	117
Scenario 2	4	12	19	26	34	8	15	25	33	44
Scenario 3	3	9	15	20	26	22	41	67	88	117
Scenario 4	0	0	0	0	0	9	16	27	35	47

NOTES: Mean growth rates used in the projections are drawn from Chapter Three of this document. These results are also displayed in Figures B.1–B.4 in Appendix B. Scenario 1: Current Cost FTR/Current Productivity; Scenario 2: South Korea's Cost FTR/Current Productivity; Scenario 3: South Korea's Productivity FTR/Current Cost per FTR; Scenario 4: Quality adjustment, according to McKinsey Global Institute (2005). With U.S. statistics serving as the base, China's factor is 0.12 and India's is 0.30.

lio of R&D investment (and, thus, patents and publications in absolute and relative importance worldwide), nor does it include potential extraordinary events that might threaten the targets set by both countries.

Table 4.4
Composition of GERD, by Source of Funding, China and India,
2008–2025 (2008 PPP US$ millions)

	GERD	BERD	HERD	GOVERD	PNPERD
India					
2010	69,900	51,554	8,708	7,998	1,858
2015	114,933	84,767	14,319	13,150	3,055
2020	151,183	111,503	18,835	17,298	4,018
2025	198,868	146,673	24,776	22,754	5,286
China					
2010	163,264	120,414	20,340	18,680	4,339
2015	269,405	198,697	33,564	30,825	7,160
2020	355,638	262,297	44,307	40,692	9,452
2025	469,475	346,257	58,489	53,717	12,478

A Qualitative Look at China's and India's Innovation

There are some qualitative aspects of innovation that are not fully reflected in the "classic" indicators of innovation. Some of these aspects relate to changes that China and India are making, and some relate to changes in the dynamics of the innovation process. The structure of business relationships has changed tremendously in the past decade, with increasing fragmentation in the production process and increasing diversity of business models, involving a multitude of players who are distributed worldwide and reallocation of capital flows around the globe. India, China, and Brazil, among others, have benefited from this trend, which is changing the innovation markets as a consequence.

"Reverse Innovation" and the Bottom of the Pyramid

In the past few decades, innovation has taken a new form. Instead of adapting to the models that proved to be successful in the developed world, entrepreneurs in emerging economies are changing the nature of the business. They are introducing innovations with the aim of reaching the huge market represented by China and India. When marketing to low-income customers, the income of customers is less significant than their collective spending power.[14] Low per capita income squeezes the profit margins, thus the companies are counting on volume to compensate. For example, the cash flows

---

[14] China's and India's lowest-income households have an annual income of about $691 billion and $378 billion, respectively (Anderson and Markides, 2007).

of emerging countries' low-income customers follow a different path; most of these customers are paid daily rather than weekly or monthly, and their income is much more unstable than in the developed world, making them acutely sensitive to price (Anderson and Markides, 2007).

The deep differences between developed and emerging world markets have challenged Indian engineers to reinvent products and cut costs and have fostered innovation in distribution, commercialization, and marketing chains. This phenomenon, known as "reverse innovation,"[15] is forcing restructuring of business by multinationals and by well-established Indian corporations (Govindarajan and Trimble, 2009).[16]

There are many examples of Indian innovations that are designed specifically for a massive, low-income market: a wood-burning stove at US\$23, a US\$34 water purification system, a US\$70 refrigerator that runs with batteries, a baby's heart monitoring system at 10 percent of the world price, etc. Some of these Indian innovations may end up in developed markets. For example, Tata Motors plans to introduce its Nano automobile to Europe, albeit with improved interiors and new safety features. Though the European version will be more expensive than the Indian one, it will be significantly cheaper than alternatives in Europe.

Chinese entrepreneurs have had similar successful experiences. For instance, Galanz, a former textile and garment manufacturer, started its "microwaves for the masses" endeavor in 1992: It now has a 35 percent world share in the microwave industry. It competes via innovation (a small, simple, energy-efficient product) instead of China's traditional cheap labor strategy (Anderson and Markides, 2007).

Reverse innovations extend to the marketing and distribution stages. As one business strategist puts it, "Companies need to learn to create markets" (Donohue, 2009). Although the objective is always the same—to reach consumers—marketing and distribution chains need to adapt to low purchasing power. Reaching the "bottom of the pyramid" requires different strategies than the ones used in the developed world. Lack of awareness of these elements has turned carefully "invented products" into failure.[17] Successful distribution models now need to be based on rural self-help groups and micro lenders plugged into villages.

This type of innovation, low priced and based in low R&D costs, is changing the dynamic of innovation and business in the world and makes forecasting S&T indicators more complex (Bellman, 2009).

---

[15] A reverse innovation, very simply, is any innovation likely to be adopted first in the developing world.

[16] India's biggest corporate group, Tata Group, is opening divisions for low-income consumers. Unilever and General Electric are among the multinationals that are following the same steps.

[17] Procter and Gamble had a huge commercial failure with a water purification system named PUR. Though the product seemed perfectly suited to the market in terms of price and the needs of India's rural population, the marketing strategy was not appropriate to make it succeed (Simanis, 2009).

High-End Innovation

Classically associated with low-level IT work, Indian and Chinese companies have climbed the value chain to become R&D providers in more complex areas. India is rapidly becoming the next global center of research, design, and innovation for many industries, including pharmaceutical (drug discovery, specialty pharmaceuticals, biologics, high value, bulk and advanced intermediate manufacturing); aerospace (in-flight entertainment, airline seat design, collision control/navigation; control systems, fuel inverting controls, first-class cabin design); consumer appliances (semiconductors, washing machines, dryers, refrigerators, digital TV, cell phones), and motorized vehicles (automobiles, tractors, locomotive) (Wadhwa, 2009b).

China is already the world's biggest exporter of computers, telecom equipment, and other high-tech electronics and is making rapid progress in infrastructure and technology. China will soon be an export powerhouse in industries such as semiconductors, passenger cars, and specialty chemicals, and, in 10 to 15 years, commercial airplanes. But the country faces challenges, including a poor reputation for protecting intellectual property rights that has prevented multinationals from transferring technology to China. China's relatively poor human capital threatens both innovation and profitability in some industries because final products cannot compete in the market due to low quality (e.g., chips).

India, on the other hand, has an innovation market that is running at high speed as a consequence of trends that favor domestic investment in high-value tasks. The positive trends are (Wadhwa, 2008) as follows:

1. an increase in education level, productivity, and quality of the workforce as a consequence of companies' investment in workforce education
2. a decrease in attrition rates due to companies' internal promotion strategies (tied to the improvement in quality) that give incentives for workers to stay and add to their human capital
3. an increase in global competitiveness due to the devaluation of the rupee
4. outsourcing to India due to changing business models in developed countries
5. recent U.S. immigration policy, which has encouraged skilled workers to return or stay in India
6. the improvement in quality and quantity of engineers: In 2004, India graduated only 125,000 bachelors in engineering, and the number doubled by 2007 and is expected to reach a half-million by 2011.

In some industries, India and China are making huge progress, particularly in the highest-value segments of global value chains.[18] For example, in the pharmaceuti-

---

[18] In the lower-value segments, such as preclinical testing, animal experimentation, and manufacturing, Chinese firms are more prevalent.

cal industry, several new business models have emerged: original proprietary research in which domestic firms provide the basic research and team up with a multinational company to advance the entire clinical-trial process and market worldwide; research partnerships; contract research organizations (CROs) that develop specific stages of the drug discovery process; and generics—India has an edge in the development of generics at lower costs.

Other Approximations to R&D

Patents tell only part of the story. Another way to approximate R&D is to follow the flow of investments. There are many examples of multinationals setting up their R&D operations and partnering with local firms in India (e.g., Palm Pre smart phone, Amazon Kindle). Currently, IBM, Big Blue, Cisco, Adobe, Cadence, Oracle, and Microsoft are developing mainstream products in India. This trend in multinational R&D operations is parallel to the growth of Indian multinationals such as Tata (produces the Nano car) and Reva (builds an electric car factory in New York State). China has had a similar experience but with different players.

Nevertheless, both countries have a weakness relating to very scarce venture activity and to entrepreneurism. Venture capital activity is considered a necessary starting point to business.[19] Wadhwa (2009b) reports that, in the first nine months of 2008, total early stage venture capital investments in India totaled $678 million, compared with the United States' $5.2 billion. China is ahead of India in terms of startups, which are mostly based on government incentives, but it is still lagging well behind startups in the United States.[20]

We could place these factors into a second category of human capital quality that we could name "entrepreneurial" talent, which has been found to have a decisive role of entrepreneurship in economic growth (Murphy, Shleifer, and Vishny, 1991). Murphy, Shleifer, and Vishny's 1991 study provided evidence that countries that reward more entrepreneurship activities than rent-seeking individuals grow faster. Recent studies using panel data on entrepreneurship (e.g., Acs et al., 2005) have not only confirmed those early findings but have highlighted the role of entrepreneurship as a mechanism that facilitates the spillover of knowledge promoting economic growth. Although we do not have statistically significant information of the relevance of this factor in India and China, we provide in the text examples that suggest that this quality factor operates on India's side. Moreover, India has not only the advantage of language (English) but also a long tradition of young people doing graduate studies abroad, especially in

---

[19] Qualitative analysis suggests that India has a slight advantage over China in venture capital activity, as Indian firms appear to be more attractive to U.S.-educated and -trained scientists and engineers.

[20] Google has launched a $100 million startup incubator that is intended to stimulate startups in the mobile IT sector.

the United States. Unfortunately, we do not have enough evidence to assess the impact of this factor on future trends.

Conclusions

The evidence presented in this chapter suggests that performance in science, technology, and innovation has continued to strengthen in recent years worldwide, but especially in China and India. China and India have become major destinations for foreign direct investment (FDI).[21] They are increasing the share of high-technology products and services in their export structures.

Both countries understand that S&T plays a crucial role in achieving economic growth. Consequently, in their five-year plans, they have set levels of R&D expenditure similar to those of industrialized countries. They have also set other measures and policy instruments in support of S&T. Plans include policy instruments aimed to improve the environment for R&D investment (legal environment for innovation, such as protecting innovation rights), foster linkages between the business and the higher education sectors, and focus on areas of research that represent urgent need for the countries' population (such as environment, biotechnology, and agriculture, among others).

However, raising R&D intensity to a level close to that of OECD countries is a challenge that requires several changes. These include, for example, migration from experimental to basic research; a more prominent role for the business sector as a performer of R&D (in both countries but eventually more in India); more connection between business and the higher education sector, which is necessary for successful innovation systems; and an increase in the quality of education to support both basic and applied research.

The higher education sector, as the main producer of intermediate outputs (including graduates in S&E), is also facing challenges in terms of the quantity and the quality of the human resources it produces. The challenge is immense; the expansion of GERD is projected to levels that will demand important increases in the number of graduates.

From our point of view, despite China's success to date, there is a mismatch between its investment in S&T and its innovation impact, as proxied by the number of patents and publications. The opinion of the business sector expressed at the Global Economic Forum confirms the difficulties in creating an innovative environment in China (Global Economic Forum, no date). Although entrepreneurs have highlighted difficulties associated with the institutional organizations for both countries as impor-

---

[21] China has also become a major source of foreign investment and, thereby, an indirect contributor to technology and innovation in some foreign companies in which it has invested (see Wolf et al., forthcoming).

tant barriers to investment, China appears to be lagging behind not only in efficiency but also in lack of competitiveness of its human capital. While India faces its own challenges, mainly related to the financial sector, its strong presence at the high end of the value chain combined with innovations addressed to low-income consumers make it likely to become a more important player in S&T in the coming decades.

Finally, our study highlights the crucial need for better statistics. This includes the harmonization of quality standards and definitions at cross-country levels. Further, important information on R&D activities is not collected or is highly aggregated and not consistent with international definitions. For instance, information is incomplete about the activities of multinational companies in China that affect supply of and demand for HRST. Harmonization of concepts and systematic collection of uncollected information will facilitate international comparisons and enable a more accurate analysis of current and future trends.

CHAPTER FIVE

Chinese and Indian Defense and Defense Procurement Spending to 2025[1]

As discussed in previous chapters, just as India's and China's endowments—vast populations, growing and increasingly sophisticated and diversified economies, and growing technology sectors—contribute to their growing strategic importance, so too do their substantial and growing defense capabilities. Accordingly, understanding the outlook for Indian and Chinese long-term defense and defense procurement is a crucial dimension of our comparative assessment of the two countries. In this chapter, we summarize our comparative analysis and forecasts of Chinese and Indian defense and defense procurement spending through 2025.

The main findings can be briefly stated as follows:

- The most likely outcome in 2025 is that Chinese defense spending will continue to exceed that of India, and the ratio of Chinese to Indian spending is likely to continue in China's favor, or even grow.
- Chinese defense procurement spending also appears likely to continue to exceed that of India, but it is plausible that the ratio of Chinese to Indian spending could either grow or shrink.
- India's defense and defense procurement spending are more transparent than China's, because of the public and detailed nature of India's public budgeting process. For its part, China's policy of reporting its defense and defense procurement spending in terms of budgetary aggregates probably omits substantial amounts of defense-related spending.
- It is easier to establish a rough floor on Chinese spending than a ceiling. Official U.S. estimates of Chinese defense spending are as much as two to three times higher than China's official estimate of defense expenditures.
- Under many—perhaps most—plausible circumstances, the recently observed high levels of double-digit growth in defense and defense procurement spending

---

[1] The authors wish to thank Deba Mohanty, Senior Fellow in Security Studies at the Observer Research Foundation (ORF), New Delhi, India; Professor Shaoguang Wang of the Chinese University of Hong Kong; and Chaoling Feng, a Doctoral Fellow in the Pardee RAND Graduate School, for their invaluable assistance.

in both China and India will likely be politically unsustainable well before 2025. It will therefore be important to carefully monitor the Chinese and Indian leadership discourse for signs that growth in defense spending will be tapering off.

- Indeed, China's recent announcement that defense spending in 2010 will grow by only 7.5 percent—about half the rate of growth of the previous year—and India's announcement that defense growth in 2010–2011 will be about 4 percent—compared with more than 20 percent growth in the previous year—suggest that both countries' defense budgets already are under pressure as a result of competing domestic demands.
- Many other international and domestic "wild cards" could affect the future trajectory of Indian and Chinese defense and defense procurement spending; a detailed analysis of these factors would be a useful complement to the present effort.

This chapter is divided into four parts. The first part describes the approach used for our analyses and forecasts. The second part reviews, summarizes, and critiques official unclassified sources of data on defense and defense procurement spending for both India and China, as well as data from other studies dealing with defense spending by the two countries. This review concludes with a set of baseline estimates of Chinese and Indian defense and defense procurement spending in 2009. The third part of the chapter presents a range of comparative forecasts of Chinese and Indian defense and defense procurement spending from the 2009 baseline figures through 2025 with varying assumptions about growth, GDP share of defense, and other factors. We close with comparative observations about the two countries, based on our analyses and forecasts, and conclusions.

Appendix C contains tables with detailed historical data on GDP and defense and defense procurement spending, estimates of nominal and real growth in these aggregates, implicit deflators used to convert spending estimates from nominal to real growth rates, exchange rates, and other technical assumptions related to the analyses and forecasts.

Analytic Approach

To understand the composition of Chinese and Indian defense spending, the analysis begins with a review of scholarly efforts that have analyzed Chinese and Indian defense spending.[2]

---

[2]   Among the more noteworthy efforts examining Chinese defense spending are Wolf et al. (1995); Jane's (1995); Wang (1996, 1999); Bitzinger (2003); Crane et al. (2005); International Institute for Strategic Studies (IISS) (2006, 2010a, especially pp. 391–392); Surry (2007); Blasko et al. (undated); and GlobalSecurity.org (2010). On the Indian defense budget, see Ghosh (2009, p. 7); Mohanty (2009, 2010); and Behera (2010).

The main line of our research was based on a review of openly available English-language primary and secondary sources, especially official documents and public statements, with modest reliance on Chinese-language sources.[3]

Historical GDP and defense and defense procurement spending data through 2009 were compiled in local currency units (LCUs) in current prices, i.e., Chinese renminbi (RMB) or Indian rupees (INR),[4] and historical real growth rates were estimated using implicit price deflators from the International Monetary Fund.[5]

We produced several dozen alternative forecasts of defense and defense procurement spending in 2025, with the estimated 2009 levels of defense and defense procurement spending and GDP serving as baselines for all of these forecasts.

Our "defense growth rate" forecasts of defense and defense procurement spending were predicated on alternative assumptions about the future average annual real growth in defense and defense procurement expenditures. For example, we used recent historical trends in the annual average real growth rates of defense and defense procurement spending and GDP through 2009 as the basis for some of our forecasts of defense and defense procurement spending levels from their 2009 baselines through 2025. To bound these forecasts, we also generated alternative forecasts assuming plausibly lower and higher real growth rates than the actual ones that have recently occurred.

An alternative set of "parametric" forecasts of defense spending relied on different combinations of assumptions about the average annual real growth rates in Chinese and Indian GDP and their corresponding defense shares of GDP.

Thus, within each of these two types of forecast—"defense growth rate" and "parametric"—we used varying assumptions to explore the wider range of plausible trajectories in defense and defense procurement spending to 2025.

Finally, to directly compare Chinese and Indian spending, we converted the defense spending and defense procurement spending estimates from constant 2009 local currency units—INR or RMB—to constant 2009 U.S. dollars using both market exchange rates (MXR) and purchasing power parity (PPP).[6]

---

[3]   For China, this included official announcements of defense spending levels for the next year, China's biannual white paper on defense, and other sources. For India, we relied on budget documentation related to India's Union Budget.

[4]   Indian budget data generally are reported in crores (tens of millions) of rupees, which necessitated conversion to billions of rupees; Chinese budget and economic data generally are reported in billions of renminbi.

[5]   See International Monetary Fund (2010b). Table 5A.1 (in International Monetary Fund, 2010b) presents the IMF's estimates of 2009 GDP for China and India in local currency units and U.S. dollars using the World Bank's 2009 average market and PPP-based exchange rates and shows that China's 2009 GDP was 2-1/2 to four times larger than that of India, depending on the exchange rate used (MXR or PPP).

Appendix C, Tables C.2 through C.4, detail the basis for our estimates of nominal and real growth in GDP and defense and defense procurement spending.

[6]   We converted constant 2009 local currency units to constant 2009 U.S. dollars using the World Bank's estimates of MXR and PPP for 2009, from World Bank (undated). The market exchange rates that we use are

Baseline Estimates for China

Crane et al. (2005) reported that "the official Chinese defense budget excludes a wide variety of military accounting items commonly included in Western budgets" and provided a "notional full Chinese military budget" consisting of the following elements:[7]

- the official Chinese defense budget
- paramilitaries (People's Armed Police, PAP)
- local support to defense and paramilitaries
- funds for foreign arms imports[8]
- defense research and development (R&D)
- defense industrial subsidies[9]
- foreign arms sales revenues.[10]

Analysts also have identified a number of additional categories of defense spending and suspected sources of additional defense resources that are not easily estimated from Chinese budget documents.[11]

6.8 RMB/$ and 48.4 INR/$, and the PPP-based rates are 3.7 RMB/$ and 16.5 INR/$, both rounded to the first decimal place. See Table 5A.1 (in International Monetary Fund, 2010b) for additional details.

[7] Crane et al. (2005, p. 133).

[8] Arms imports are generally believed to be funded from hard-currency accounts managed by the State Council and are not included in the official Chinese estimate of defense spending.

[9] As Chinese defense industry reforms and restructuring appear to have eliminated most defense industry subsidies, we follow IISS's lead and do not include these in our estimate. The IISS's *The Military Balance* included defense subsidies in 2003 and 2006, but its estimate of Chinese defense spending in 2008 dropped these items, stating: "[T]he level of state subsidies to the defence industry is now unlikely to be significant and is no longer taken into account by *The Military Balance*" (IISS, 2010a, p. 392).

[10] Wang (1999) suggests that the PLA receives commissions for arms exports as an extra-budgetary source of Chinese defense resources. As will be described, we assume that most foreign arms sales revenues go to the defense industry groups that are responsible for defense production of the arms that are sold, either directly or as a pass-through from the Ministry of Defense. We acknowledge the possibility of additional potential subsidies to defense accounts resulting from these transactions.

[11] For example, Crane et al. (2005) mention spending on nuclear weapons and strategic rocket programs and extra-budget revenue (yusuanwai), and Bitzinger and Lin (1994) report that "China's nuclear weapons program is largely hidden within the PRC's nuclear energy and space programs."

For his part, Wang (1999) mentions the following expenses as military expenditures that are detailed in non-defense budget categories: People's Armed Police; defense research, development, test, and evaluation (RDT&E); construction of research facilities and military production lines operated by civilian institutions; one-time demobilization expenses; subsidies to military production; and special appropriations for arms acquisitions from abroad. Wang also mentions commercial earnings from domestic business activities and PLA commissions for arms exports as extra-budgetary sources of Chinese defense resources.

Using a slightly modified version of Crane et al.'s framework,[12] we estimated China's total defense spending in 2009.[13]

We began with the official Chinese defense expenditure estimate of 481 billion RMB in 2009 and added state, provincial, and local funding for PAP paramilitary forces (about 75 billion RMB) and estimated arms imports in 2009 (about 10 billion RMB).[14]

We then estimated defense-related R&D and government-funded science and technology (S&T) in 2009.[15] We did this by increasing the IISS estimates for 2008 (46.1 billion RMB in defense-related R&D and 34.5 billion RMB in defense-related government-funded S&T) by the estimated growth rate in the official Chinese defense budget from 2008 to 2009 (14.9 percent). Thus, we assume that the growth in nominal defense-related R&D between 2008 and 2009 was at the same rate as the nominal growth rate in China's official defense budget during the same period. This yielded an estimated total of 53 billion RMB in defense-related R&D and 40 billion RMB in S&T spending in 2009 current prices.[16]

---

[12] As discussed below, the principal differences with Crane et al.'s framework are that, in line with more recent analyses, we have eliminated defense industry subsidies and have added the net defense-related revenues of the defense industry groups. In addition to Crane et al. (2005), we relied on the following sources: Bitzinger and Lin (1994); IISS (1996, pp. 270–275; 2006, pp. 249–253; 2009, pp. 375–376; 2010a, pp. 391–393); and Wang (1999, pp. 334–349).

[13] It is noteworthy that the Chinese also appear to distinguish between "defense spending," which is captured entirely in the official defense budget, and "military spending, which also includes additional items of defense-related expenditure. According to press describing a recent internal Chinese report,

> The gap [between the official estimate of Chinese defense spending and the actual level of defense-related spending] shows the PLA appears to have a concept of "military spending," which is different from—and larger than—a defense budget. The sources said military spending represents the defense budget plus military-related outlays for the Ministry of Industry and Information Technology and other organs under the State council. ("China's 2010 Military Spending 1.5 Times Larger Than Defense Budget," 2010)

[14] Our estimate of Chinese arms imports is based on Grimmett (2010) and Stockholm International Peace Research Institute (undated).

[15] Wang Shaoguang reported in 1999 that at that time military research and development was coordinated by the Commission on Science, Technology and Industry for National Defense (COSTIND), and that the defense portion of military R&D was to be found in the general R&D fund account called "expenditure on research" (*yanzhi jingfei*) and a counterpart in the new product development fund called "expenditure on test, evaluation and prototypes" (*shizhi jingfei*). See Wang (1999, p. 339).

[16] See IISS (2010a, p. 392). Past estimates of China's military-related R&D and S&T spending include the following: growth in Chinese military-related R&D from 1 billion RMB in 1989 to 6.9 billion RMB in 1998 and growth in test and evaluation (T&E) from 1.7 to 6.5 billion RMB over the same period (Wang, 1999); 5.0 billion RMB in direct military R&D allocations in 1993 (IISS, 1996); 6.9 billion RMB in Chinese military-related R&D in 1998 and test and evaluation amounting to 6.5 billion RMB; 23.1 billion RMB in R&D and 25.2 billion RMB in "New Product Expenditure" in 2003; 45 billion RMB in R&D and 47.8 billion RMB in "New Product Expenditure" in 2006; and 46.1 billion RMB in R&D and 34.5 billion in "government funded science and technology" in 2008. See IISS, *The Military Balance*, various years.

Our estimate of about eight billion RMB in revenues from foreign arms sales in 2009 was subtracted from our estimate of the defense share of defense industry group revenues, described next. We did not attempt to estimate the profits or commissions from foreign arms sales that might be used to advance Chinese military purposes, as most contemporary scholars appear to view these profits and commissions as modest.[17] In any event, the amounts involved have a negligible effect on the bottom-line estimates for defense spending.

Our estimate differs from most others in one other important respect: our handling of Defense Industry Group (DIG) revenues. China's ten DIGs are state-owned enterprises responsible for the indigenous production of Chinese defense goods and, increasingly, a range of civilian goods such as commercial ships, aircraft, and electronics as well. In theory, DIG revenues related to defense production should approximately equal the total of the "Equipment" account in the official budget, plus foreign arms sales of defense goods produced by the DIGs. Building on Surry's earlier effort to estimate the defense-related revenues of China's DIGs (Surry, 2007), we estimated total DIG revenues for 2009 by projecting 14.9 percent nominal growth over the estimated 2008 level of 909 billion RMB, or 1,044 billion RMB. Using Surry's 2007 estimate that up to 35 percent of defense industry group revenues was defense-related and evidence that this share has been declining, we then assumed three alternative defense shares of total DIG revenues: a "low" estimate of 20 percent, a "mid" estimate of 25 percent, and a "high" estimate of 30 percent. We then subtracted from the resulting estimates of defense-related revenues the official Chinese estimate for the "Equipment" account of defense spending and estimated foreign arms sales.[18] This yielded an estimated 41 to 145 billion RMB in revenues, depending on whether the defense share of DIG revenues was assumed to be 20, 25, or 30 percent of total revenues (see Table 5.1).[19]

That all of the estimates of defense-related DIG revenues exceeded the sum of the "Equipment" account and estimated foreign arms sales suggests that the difference may be a form of off-budget defense spending. As a practical matter, we used our middle estimate of defense-related DIG revenues in 2009 for our baseline estimates

---

[17]   See, for example, Wang (1999), who cautions against exaggerating PLA profits from arms sales. If PLA commissions on foreign arms sales constituted 10 percent of the total revenues, the profits from our estimated 8.0 billion RMB in 2009 arms sales would be less than 1 billion RMB—a tiny share of our estimate of 700–800 billion RMB in total Chinese defense spending in 2009. Thus, the inclusion or omission of profits is unlikely to affect our overall estimates of Chinese defense spending.

[18]   The revenue for foreign arms sales goes primarily to the defense industry groups, although some revenues reportedly also go to the PLA. See Crane et al. (2005, p. 131) and IISS (2006, p. 252).

[19]   See Appendix C, Table C.5, for our detailed estimates of Defense Industry Group revenues and defense shares.

Table 5.1
Assumptions Behind Estimates of Net Defense Industry Group Revenues in 2009 (billions of RMB, current prices)

	Low	Mid	High
Estimated total DIG revenues	1,044	1,044	1,044
Estimated total defense-related DIG revenues	209	261	313
Official Chinese estimate: "Equipment"	160	160	160
Estimated foreign arms sales	8	8	8
Net DIG revenues	41	93	145

NOTES: As described in the main text, Surry (2007) estimated that the recent defense share of DIG revenues could have been as high as 35 percent, but this share appears to be declining. Our low estimate assumes that 20 percent of the total revenues of DIGs are defense-related; our middle estimate assumes that 25 percent of revenues are defense-related; and our high estimate assumes that 30 percent of revenues are defense-related. See Appendix C for historical estimates of DIG revenues.

and forecasts and simply note that there is some uncertainty about the true value of those revenues.[20]

As just discussed, our estimate of total Chinese defense spending in 2009 is based on the official estimate of 481 billion RMB, plus spending on the PAP (75 billion RMB); foreign arms purchases (10 billion RMB); defense research and development (53 billion RMB); government-funded defense science and technology (40 billion RMB); and net DIG revenues of 41, 93, or 145 billion RMB, the number depending on whether the defense share of DIG revenues is assumed to be 20, 25, or 30 percent. Based on these estimates, our middle estimate of Chinese defense spending in 2009 is 752 billion RMB, with a range of plus or minus 52 billion RMB (about 6.9 percent), or 700 to 804 billion RMB, depending on the assumed defense share of DIG revenues.

Our estimates are between 2.1 and 2.4 percent of Chinese GDP in 2009,[21] higher than the official Chinese estimate of 1.38 percent of GDP and generally in line with other studies.[22] However, these other estimates do not include net defense industry

---

[20] Improving the basis for estimating defense-related DIG revenues is clearly an area deserving additional analysis.

[21] GDP share is based on the International Monetary Fund (2010b) estimate of Chinese GDP in 2009 of 34,050.7 billion RMB.

[22] Crane et al. (2005) estimated the defense share of Chinese GDP in 2003 to be between 2.3 and 2.8 percent of GDP; IISS (2010a) estimated 2008 Chinese defense spending at 577.8 billion RMB or 1.88 percent of GDP; and Stockholm International Peace Research Institute (undated[a]) estimated 2009 Chinese defense spending at 686 billion RMB, or about 2.05 percent of GDP.

group revenues, but they are reasonably close to those in recent press reporting on Chinese defense and military spending levels.[23]

In estimating Chinese defense procurement, we include the official Chinese estimate of spending on "Equipment," estimated foreign arms purchases, and net defense industry group revenues, which leads to an estimate of 211 to 315 billion RMB in defense procurement spending, or 30–40 percent of total defense spending (see Table 5.2).

As just described, because of the many gaps and uncertainties in China's defense procurement and defense spending data, our baseline estimates are necessarily partial; we have included only those elements we could estimate with some confidence. Put another way, these estimates generally should be viewed as floors on Chinese defense procurement and defense spending.

Baseline Estimates for India

India's annual Union Budget is detailed in a large number of "Demands for Grants" (DGs) that contain the government of India's requested budget levels for its various ministries, departments, and other governmental activities, as well as updated estimates of spending in prior years.[24] In India's budgetary parlance, "Demands for Grants" correspond generally to what in U.S. terminology would be the Department of Defense part of the budget request submitted to the Congress by the president.

Table 5.2
Assumptions Behind Estimates of Chinese Defense Procurement Spending in 2009 (billions of RMB, current prices)

	Low	Mid	High
Official Chinese estimate: "Equipment"	160	160	160
Foreign arms purchases	10	10	10
Net DIG revenues	41	93	145
Total	211	263	315

NOTES: See the main text and the notes to Table 5.1 for an explanation of the calculation of net DIG revenues. Estimates of Chinese foreign arms purchases are based on Grimmett (2010) and Stockholm International Peace Research Institute (2010a).

---

[23] According to press reporting, a recent internal PLA study estimated Chinese "military spending"—including both the official defense budget and other defense-related spending by the State Council—at 788 billion RMB in 2010, or about 2.5 percent of GDP. This is very close to our estimates of Chinese defense-related spending in 2009. See "China's 2010 Military Spending 1.5 Times Larger Than Defense Budget" (2010).

[24] The various Demands for Grants associated with India's Union Budget for 2009–2010 can be found at Government of India (undated[b]).

Table 5.3 presents recent budgets and spending levels for the eight "Demands for Grants" that are associated with India's Ministry of Defence and that constitute India's spending on national defense.

As shown in the table, total planned Indian defense spending for 2009–2010 was 1,666.6 billion rupees, a nominal increase of 21.4 percent over the 2008–2009 defense spending level.

Also shown in the table, the top three defense requirements for the 2009–2010 budget year were Capital Outlays, the Indian Army, and Pensions, accounting for about three-fourths of total estimated spending by the Ministry of Defence that year. Moreover, spending on the Indian Army was more than four times the spending on the Indian Air Force and seven times the spending on the Indian Navy. Whereas Capital Outlays make up almost one-third of total defense spending, research and development make up less than one-thirtieth of the total defense budget.

Because India does not consider spending on paramilitary forces to be part of its spending on national defense, we do not include this spending by the Indian Ministry of Home Affairs in our overall estimate of Indian defense spending. The reader will recall, however, that the Chinese do consider its paramilitary force—the People's Armed Police—to be part of its national defense, and we accordingly include spending on the PAP in our defense spending estimate for China. What difference would the inclusion of Indian spending on paramilitary forces make to our overall defense spending estimates for India?

Table 5.3
Indian Ministry of Defence Demands for Grants, 2009–2010 (billions of INR, current prices)

Demand for Grants	2008–2009 Budget	2008–2009 Revised	2009–2010 Budget
DG20. Ministry of Defence	23.7	23.9	31.7
DG21. Defence Pensions	155.6	202.3	217.9
DG22. Defence Services—Army	362.7	482.0	586.5
DG23. Defence Services—Navy	74.2	80.3	83.2
DG24. Defence Services—Air Force	108.6	122.0	143.2
DG25. Defence Services—Defence Ordnance Factories	3.5	13.3	8.3
DG26. Defence Services—Research and Development	33.9	38.4	47.6
DG27. Capital Outlays on Defence Services	480.1	410.0	548.2
Grand total	1,242.3	1,372.2	1,666.6

SOURCE: Government of India (undated[b], Part II).

As described in Table 5.4, we estimate Indian spending on the Ministry of Home Affairs' paramilitary forces in 2009–2010 to be about 230 billion INR. Thus, if paramilitary-related spending is added to Indian defense spending, the total would be nearly 1.9 trillion INR (1,667 billion INR for defense plus 230 billion INR for paramilitary forces), or about 13.8 percent higher than our estimate of Indian defense spending alone. Thus, although the inclusion of Indian spending on paramilitaries would enhance conceptual comparability and narrow the gap between the two countries' defense-related spending, we have chosen to rely on each country's policy regard-

Table 5.4
Other Defense-Related Indian Ministerial Demands for Grants, 2009–2010,
(billions of INR, current prices)

Demand for Grants	2008–2009 Budget	2008–2009 Revised	2009–2010 Budget
Ministry of Home Affairs Paramilitary			
51. Ministry of Home Affairs			
51.5 Intelligence Bureau	5.3	6.8	7.9
51.7 Civil Defence	0.5	0.3	0.7
51.8 Home Guards	0.5	0.5	0.5
52. Cabinet			
52.8 Special Protection Group	1.1	1.8	2.3
53. Police			
52.1–53.14 Total Police[a]	192.2	231.7	218.4
Total Home Affairs Paramilitary	199.6	241.0	229.7
Memo: Department of Atomic Energy[b]			
4. Atomic Energy	39.1	49.6	53.0
5. Nuclear Power Schemes	8.9	18.1	24.7
Memo: Department of Space[c]			
89. Department of Space	38.6	32.9	40.7

SOURCE: Government of India (undated[b], Part II).

NOTE: Totals reflect rounding.

[a] This amounts to approximately one-third of the total 2009–2010 budget for Demand No. 53, the Ministry of Home Affairs Police function.

[b] Indian nuclear weapon–related defense programs make up an unknown share of the Department of Atomic Energy's total spending.

[c] Indian missile-related defense programs make up an unknown share of the Department of Space's total spending.

ing the relationship between defense and its paramilitary forces.[25] Definitions aside, as a practical matter, in interpreting our forecasts, readers should keep in mind that if Indian paramilitaries are included in Indian defense spending, our forecasts for Indian defense spending may be nearly 14 percent higher.

India's spending on defense nuclear, missile, and space capabilities is not well documented, a point also emphasized in our earlier discussion of China's defense spending. According to a recent analysis:

> National defence accounts for 13 percent of the central government expenditure and if military-and-security-related components of other ministries or departments like the Ministry of Home Affairs, Department of Atomic Energy and Department of Space are taken into consideration, the larger national security considerations actually consume more than 20 percent of total expenditure.[26]

Thus, our analysis of national security–related spending by other ministries included a review of spending by India's Departments of Atomic Energy and Space. Table 5.4 describes the 2009–2010 budgets of India's Department of Atomic Energy, which is responsible for the development of Indian nuclear weapons and missilery, and spending by the Department of Space, which is responsible for other defense-related space capabilities. The lack of explicit treatment of defense nuclear, missile, and space activities in Indian budget documents makes it almost impossible to estimate Indian spending on nuclear weapon and defense space programs from the budget documents.

Compared with India's roughly 1.7 trillion rupees for defense spending and the Ministry of Home Affairs' 230 billion rupees for paramilitary capabilities, the budgets of these departments—less than 100 billion rupees for the Department of Atomic Energy and 40 billion rupees for the Department of Space—are quite small relative to overall Indian national security spending. The small amounts associated with the Departments of Atomic Energy and Space may mean that some of the costs related to nuclear weapons and missiles and defense space activities are hidden elsewhere in the Indian budget. We conclude from this analysis of available data that Indian spending on defense, nuclear, missile, and space capabilities remains as much of a puzzle as Chinese spending on these capabilities: In neither case are we able to provide a reliable estimate.

---

[25] We acknowledge that an argument can be made in favor of an "apples-to-apples" comparison with China's defense spending, in which Indian spending on paramilitary forces is also included so that it is more comparable to our estimate of Chinese defense spending, which includes spending on the People's Armed Police. This would obviously help to reduce the gap between Chinese and Indian defense spending that favors China.

[26] See Mohanty (2010), pp. 15–16.

Baseline Estimate of Indian Defense Procurement Spending

Of the Indian Ministry of Defence's eight Demands for Grants, three—related to Capital Outlays, Coast Guard Organisation, and Defence Ordnance Factories—are directly related to procurement (see Table 5.5)

Table 5.5
Composition of Indian Defense Procurement, as of 2009–2010 Budget
(billions of INR, current prices)

Demand for Grants	2008–2009 Buget	2008–2009 Revised	2009–2010 Budget
27.3. Aircraft and Aero-Engine	146.6	134.2	153.1
Army	4.3	4.0	10.2
Navy	22.5	21.5	24.5
Air Force	119.9	108.7	118.4
27.4. Heavy and Medium Vehicles	13.0	13.1	9.3
Army	12.9	12.1	8.3
Navy	0.1	0.1	0.1
Air Force	0.0	0.9	0.8
27.5. Other Equipment	158.0	129.4	191.1
Army	83.5	62.7	111.2
Navy	11.7	15.2	11.0
Air Force	62.9	51.5	68.9
27.6. Naval Fleet	72.4	40.0	68.4
27.13. Procurement of Rolling Stock	1.1	1.3	1.7
Demand No. 27 total	391.2	318.0	423.6
20. Ministry of Defence			
20.3.4047 Coast Guard Organisation[a]	9.5	7.0	13.0
25. Defence Ordnance Factories	3.5	13.3	8.3
Grand total	404.2	338.3	444.9

SOURCE: Government of India (undated[b], Part II).

NOTE: The numbers preceding the item description are the numbers of the Demand for Grants and subtitle.

[a] This is a capital outlay account that includes acquisition of ships, fleets, aircraft, and major works for the Coast Guard Organization. It probably also includes some nonprocurement outlays on such capital projects as docks and berths.

As shown in the table, we estimate total Indian defense procurement spending in 2009–2010 at about 445 billion rupees, or a little over one-quarter of overall Indian defense spending.[27]

In addition to these procurement-related accounts, a fourth Demand for Grants—Research and Development (No. 26)—also contributes to the defense research, development, and acquisition functions—with spending pegged at nearly 48 billion INR in 2009–2010.

Finally, as described above, it is not clear where procurement related to defense nuclear and missile programs takes place, e.g., whether it is conducted within the Ministry of Defence or by the Departments of Atomic Energy and Space. However, the defense-related spending of these departments is nowhere elaborated and remains a key uncertainty in estimating overall Indian defense procurement spending.

Using the baseline Chinese and Indian levels just described, we now provide our estimates of Chinese and Indian defense and defense procurement spending in 2025.

Comparative Forecasts of Defense and Defense Procurement Spending in 2025

We conducted two types of forecasts of Chinese and Indian defense spending in 2025, both of which used the estimates of the 2009 baseline levels of spending, described above, as a starting point:

- The first, which we call "defense growth rate" forecasts, projected real growth in defense spending at various constant average annual real growth rates from 2010 to 2025.
- The second, which we call "parametric" forecasts, assumed several alternative annual real growth rates in GDP and several alternative GDP shares for defense spending.

Both types of forecasts are described below.

Growth Rate Forecasts of Defense Spending in 2025

Using estimated defense spending levels in 2009, the mean real GDP growth rates reported in Chapter Three, and recent trends in real Chinese and Indian defense spending and real GDP growth over the last decade, we posited several plausible alternative

---

[27] Indian arms purchases on the international market have been estimated to account for 70 percent or more of Indian defense procurement. See IISS (2010b, p. 473) and Joseph (2010).

Indian international arms purchases appear to be included in India's defense procurement accounts and are managed by the Defence Offset Facilitation Agency under the Department of Defence Production. See "India's Arms Imports to Touch $30 bn by 2012: Assocham" (2010).

rates of real growth in these defense aggregates and generated a wide range of alternative forecasts based on these different growth rate assumptions. Each forecast assumes a different annual growth rate that is sustained throughout the 2010–2025 period.[28]

As described above, between our higher and lower estimates, the middle estimate of Chinese defense spending in 2009 is 752 billion RMB, which translates to 110.1 billion U.S. dollars (MXR), or 201.1 billion U.S. dollars (PPP). The estimate of Indian defense spending in 2009 is 1,666.6 billion INR, which translates to 34.4 billion U.S. dollars (MXR), or 101.2 billion U.S. dollars (PPP). Thus, the ratio of Chinese to Indian defense spending in 2009 is estimated at 2.0 to 3.2, depending on whether PPP-based or market exchange rates are used. Applying PPP conversion rates raises the Indian estimate relative to that of China.

Table 5.6 presents a range of forecasts based on different assumptions about the average annual real growth in Chinese and Indian defense spending from 2010 to 2025:

- The "GDP Meta-Analysis" forecasts in Table 5.6 assume that Chinese and Indian defense spending will grow at the respective mean GDP annual growth rates from the meta-analysis of GDP forecasts reported in Chapter Three, i.e., 5.7 percent for China and 5.6 percent for India.
- The "Historical GDP" forecasts in the table assume that Chinese and Indian defense spending will grow at their respective annual average real GDP growth rates observed in the decade to 2009, i.e., 10.3 percent for China and 6.9 percent for India.[29]
- The "Historical Defense" forecasts in the table assume a continuation of defense spending growth at the annual average real growth rate in defense spending in the decade to 2009, i.e., 12.1 percent for China and 6.5 percent for India.[30]

As shown in the ratios of Chinese to Indian defense spending at the bottom of the table, because China and India are posited to grow at comparable rates, the "GDP Meta-Analysis" would maintain China's current advantage in defense spending, whereas the continuation of recent trends captured in the "Historical GDP" and "Historical Defense" cases suggests that the gap between the forecasts for Chinese and Indian defense spending would widen.

The results in Table 5.6 suggest that if Indian defense spending grew at the higher rate of annual historical GDP real growth (6.9 percent), and if China's defense spend-

---

[28] The assumption is admittedly arbitrary but, in the interest of simplicity, it is reasonable for the purpose of the comparative forecasts.

[29] See Appendix C, Table C.2, for estimates of historical Chinese and Indian GDP and real GDP growth through 2009.

[30] See Appendix C, Table C.3, for estimates of historical Chinese and Indian defense spending and real growth through 2009.

Table 5.6
Growth Rate Forecasts of Chinese and Indian Defense Spending in 2025 (in billions)

	Defense 2009	GDP Meta-Analysis	Historical GDP	Historical Defense
China				
Growth rate, %		5.7	10.3	12.1
Constant 2009 RMB	752.0	1,825.7	3,609.3	4,703.1
Constant 2009 U.S. $ (MXR)	110.1	267.2	528.3	688.5
Constant 2009 U.S. $ (PPP)	201.1	488.1	965.0	1,257.5
India				
Growth rate, %		5.6	6.9	6.5
Constant 2009 INR	1,666.6	3,985.2	4,847.0	4,565.3
Constant 2009 U.S. $ (MXR)	34.4	82.3	100.1	94.3
Constant 2009 U.S. $ (PPP)	101.2	242.0	294.3	277.2
China-to-India ratio in 2025				
Constant 2009 U.S. $ (MXR)	3.2	3.2	5.3	7.3
Constant 2009 U.S. $ (PPP)	2.0	2.0	3.3	4.5

NOTES: Defense values for 2009 are estimated. GDP Meta-Analysis forecasts are based on the mean growth rates from the meta-analyses reported in Chapter Three: 5.7 percent for China and 5.6 percent for India. Historical GDP forecasts are based on a continuation of the average annual real growth in GDP between 2000 and 2009 out to 2025: 10.3 percent for China and 6.9 percent for India. Historical Defense forecasts are based on a continuation of the average annual real growth in defense spending between 2000 and 2009 out to 2025: 12.1 percent for China and 6.5 percent for India. World Bank LCU–U.S. dollar exchange rates were used: China-PPP, 3.7; China-MXR, 6.8; India-PPP, 16.5; and India-MXR, 48.4. See Chapter Three and the tables in Appendix C for additional details on the basis for these alternative real average annual growth rates. We note that the Historical Defense forecast is based on our estimate of the real growth rate in the official Chinese estimate of "defense spending," which, as described above, does not include some other categories of "military spending."

ing grew at the lower rate of GDP growth (5.7 percent) shown in the meta-analysis column, India could whittle down China's defense spending advantage to a ratio of about 2.6. Judging by the recent history of real growth in defense spending for the two countries, this combination of circumstances seems implausible but not inconceivable.

Parametric Defense Spending Forecasts Based on Varied Assumptions About Real GDP Growth Rates and Defense Shares of GDP

Key uncertainties for both India and China include the average real GDP growth rates of each country through 2025 and the average percentage of GDP that policymakers in each country will be willing to devote to defense. Accordingly, we generated another set of forecasts that we call "parametric" forecasts and that revolve around different assumptions about the defense share of GDP for China and India, as well as their respective average real GDP growth rates.

Given China's current advantage over India in GDP, it is not surprising that when Chinese and Indian GDPs are assumed to grow at identical real growth rates and the defense shares of GDP are identical, China's defense spending will increase its lead over India. In these cases, the ratio of Chinese to Indian defense spending in 2025 is four times larger using market exchange rates, and 2.5 times larger using PPP-based exchange rates. Are there any combinations of Chinese and Indian GDP growth rates and defense shares of GDP in which India can reduce China's advantage in defense spending by 2025 or even overtake China?

To address this question, we generated forecasts of Chinese and Indian defense spending in 2025 using alternative assumptions that plausibly envelope the recent Chinese and Indian historical experience in average annual real GDP growth rates (5.0, 7.5, 10.0, and 12.5 percent real growth) and shares of GDP devoted to defense (2.0, 2.5, and 3.0 percent of GDP).[31] Using the same set of bounding assumptions for China and India results in 12 cases for each country (four possible GDP growth rates times three possible GDP shares), or a total of 144 paired cases.

In the case that most favors China, in which China's average real GDP growth is 12.5 percent and its defense share of GDP is 3.0 percent, although India grows at 5.0 percent and commits only 2.0 percent to defense, the ratio of Chinese to Indian defense spending can be as high as 11.4 (PPP) to 18.3 (MXR). On the other hand, only in the case where Indian GDP grows at the highest real growth rate (12.5 percent) and its defense share is 3 percent, and China's GDP grows at the lowest growth rate (5.0 percent) and its defense share is 2 percent, can Indian defense spending approach, or surpass, that of China.

To conclude this discussion of defense spending forecasts, the results strongly suggest that China's margin over India in defense spending is likely to continue or grow by 2025.

Growth Rate Forecasts of Defense Procurement Spending in 2025

Our forecasts of Chinese and Indian defense procurement spending in 2025 suggest that China is also likely to retain its advantage in the area of spending on military modernization (see Table 5.7).

Table 5.7 summarizes our forecasts of Chinese and Indian defense procurement spending using four alternative real growth rates: (1) the mean real GDP growth rates from the meta-analyses reported in Chapter Three ("GDP Meta-Analysis," 5.7 and

---

[31] We assumed a 2009 GDP of 34,050.7 billion RMB for China and 59,520.0 billion INR for India. International Monetary Fund (2010b). As described above, we estimate the defense share of Indian GDP to be about 2.9 percent in 2009, and our middle estimate for China was 2.2 percent, with a range of 2.1 to 2.4 percent.

Table 5.7
Growth Rate Forecasts of Chinese and Indian Defense Procurement Spending in 2025 (in billions)

	Defense Procurement 2009	GDP Meta-Analysis	Historical GDP	Historical Defense	Historical Procurement
China					
Growth rate, %		5.7	10.3	12.1	12.8
Constant 2009 RMB	263.0	638.5	1,262.3	1,635.5	1,817.9
Constant 2009 U.S. $ (MXR)	38.5	93.5	184.8	239.4	266.1
Constant 2009 U.S. $ (PPP)	70.3	170.7	337.5	437.3	486.1
India					
Growth rate, %		5.6	6.9	6.5	12.8
Constant 2009 INR	444.9	1,063.9	1,293.9	1218.6	3,068.8
Constant 2009 U.S. $ (MXR)	9.2	22.0	26.7	25.2	63.4
Constant 2009 U.S. $ (PPP)	27.0	64.6	78.6	74.0	186.3
China-to-India ratio					
Constant 2009 U.S. $ (MXR)	4.2	4.3	6.9	9.5	4.2
Constant 2009 U.S. $ (PPP)	2.6	2.6	4.3	5.9	2.6

NOTES: Defense Procurement 2009 (Column 1) is estimated. GDP Meta-Analysis forecasts (Column 2) are based on defense procurement growth at the mean GDP growth rates from the meta-analyses reported in Chapter Three: 5.7 percent for China and 5.6 percent for India. Historical GDP forecasts (Column 3) are based on defense procurement growth at the average annual real growth in GDP between 2000 and 2009 out to 2025: 10.3 percent for China and 6.9 percent for India. Historical Defense forecasts (Column 4) are based on growth in defense procurement spending at the average annual real growth in defense spending between 2000 and 2009 out to 2025: 12.1 percent for China and 6.5 percent for India. Historical Procurement forecasts (Column 5) are based on a continuation of the average annual real growth in defense procurement spending between 2000 and 2009 out to 2025: 12.8 percent for both China and India. See Chapter Three and the tables in Appendix C for additional details on the basis for these alternative real average annual growth rates. We note that the Historical Procurement forecast is based on our estimate of the real growth rate in the official Chinese estimate of the "Equipment" portion of "defense spending," which, as described above, does not include some other categories of military spending. As noted in the accompanying text, it is highly unlikely—for both economic and political reasons—that real rates of growth shown in Historical Procurement (Column 5) could be sustained through 2025.

5.6 percent real growth); (2) each country's historically observed real GDP growth rates over the last decade ("Historical GDP," 10.3 and 6.9 percent for China and India, respectively); (3) each country's historically observed real rate of growth in defense spending over the last decade ("Historical Defense," 12.1 and 6.5 percent); and (4) each country's historically observed real rate of growth in defense procurement spending over the last decade ("Historical Procurement," 12.8 percent). We also report the ratio of Chinese to Indian defense procurement spending in 2025 for each pair of these growth rate assumptions.

As described above, we estimate 263 billion RMB in Chinese defense procurement spending in 2009, which translates to 38.5, or 70.3 billion U.S. dollars (MXR and PPP, respectively), and 444.9 billion INR in Indian defense procurement spending in 2009, translating to 9.2, or 27.0 billion U.S. dollars (MXR and PPP, respectively). The estimated ratio of Chinese to Indian defense procurement spending in 2009 is 2.6 (PPP) or 4.2 (MXR). As noted above, employing PPP conversion rates raises the Indian estimates relative to those of China.

As shown in Table 5.7, because the rates of GDP growth are nearly identical for China and India, in the "GDP Meta-Analysis" and "Historical Procurement" forecasts, pairing them has very little effect on the ratio of Chinese to Indian defense procurement spending by 2025. The remaining two cases—"Historical GDP" and "Historical Defense"—suggest that China's spending advantage on defense procurement could widen. In only one case, where Indian defense procurement grows at the unlikely "Historical Procurement" rate of 12.8 percent and China's defense procurement grows at a rate of 5.7 percent, would India's defense procurement surpass that of China.

Our analysis of these and other cases suggests that, although it is most likely that China will enjoy a continued margin over India in defense procurement spending, it is plausible that China's margin could either grow or shrink. The results also suggest the implausibility of India surpassing China in its defense procurement spending in 2025. Such an outcome would require a combination of a continued high rate of real Indian growth in defense procurement spending coupled with a low rate of Chinese growth in its defense procurement.

Broader Comparisons

Having compared our forecasts of defense and defense procurement spending for India and China from a quantitative perspective, we now turn to a broader comparison between the two countries.

Update: The Chinese and Indian 2010 Defense Budgets

Since the completion of our analysis, in February 2010 India announced its Union Budget for 2010–2011, including its estimates of defense and defense procurement spending for the year, and in March 2010 China announced its defense spending plans for the year. What do the newest defense budget announcements suggest about future defense and defense procurement levels in each country?

Chinese Defense Spending in 2010

After 20 years of annual double-digit growth in its official estimate of nominal defense expenditures, in March 2010 China announced that total Chinese defense spending in 2010 would be 532.1 billion RMB (about 78 billion U.S. dollars) or about 1.4 percent of GDP; this level of spending reflected growth of only 7.5 percent, about half of the preceding year's announced 14.9 percent growth rate.[32] The announcement was met with both surprise and puzzlement by most observers of Chinese military spending, and its implications for the longer-term trend in Chinese defense spending remain unclear. At least one implication for the present study is clear, however: Chinese spending in 2010 is, at least in the short term, likely to be closer to our low forecasts of defense and defense procurement spending than to our middle and high estimates.

Indian Defense Spending in 2010

In February 2010 India announced its Union Budget for 2010–2011, including a budget estimate for defense of 1,473 billion INR (about 31.9 billion U.S. dollars), a modest increase of 3.98 percent over the previous year's announced planned spending level of 1,417 billion INR.[33] This reflected a significant reduction in the growth of Indian defense expenditures and also suggests that our low estimates for Indian defense growth may best capture the current trend in Indian defense spending. A diminished Indian appetite for defense spending also appears substantiated by the government of India's acceptance of the Thirteenth Finance Commission's recommendations to reduce defense's share of Indian GDP from 2.2 percent in 2009–2010 to 1.76 percent in to 2014–2015.[34] Thus, in India's case as well, recent developments suggest lower rates of growth in defense and defense procurement spending that are closer to our low estimates than to our high ones.

Although it is far too early to say with any certainty whether this is the beginning of a new trend of lower defense spending by China and India, it does give greater credence to our low estimates of future defense spending, while also highlighting the challenges of forecasting future defense spending for these two countries.

---

[32] The Chinese also announced that the actual level of defense spending in 2009 had been 482,985 billion renminbi, 102.1 percent of the budgeted figure, and a year-on-year increase of 72,844 billion renminbi, or 17.8 percent. This was higher than the 14.9 percent that was announced the previous year. See "China's Defense Budget to Grow 7.5 Percent in 2010: Spokesman" (2010); "China's Defense Spending to Increase 7.5 Pct in 2010: Draft Budget" (2010b); Moss (2010); and He (2010).

[33] The base defense budget in 2009–2010 was 1,473.44 billion INR. See Behera (2010). In February 2010, Defence Minister A. K. Antony explicitly pegged future Indian defense spending to India's economic performance, stating: "India's defence expenditure is 2.5 percent of its GDP and the economy is expected to grow at 8 to 10 percent for the next two decades. The expenditure on defence in absolute terms is also bound to increase in equal proportion." See "Defence Expenditure Increase in Proportion with Growth: India" (2010).

[34] See Government of India (2009, pp. 379–380).

India or China: Which Is Likely to Have the Edge in 2025?

Under the most plausible assumptions, Chinese defense and defense procurement spending will exceed that of India in 2025, and the ratio of Chinese to Indian spending will grow. The fact that this result depends on a comparison that favored India (because of the greater completeness of its spending data), and probably underestimated China's spending (because of data gaps), suggests that the gaps in 2025 may favor China by more than our estimates and forecasts suggest. In addition, reported ineffectiveness and inefficiencies in the Indian research, development, and acquisition system suggest that, unless India succeeds in major reforms, the gap between China and India in the production of actual defense capabilities—quantitative and qualitative—could be even larger.

Guns, Butter, and Defense Share of GDP

The rather unexpected and dramatic reduction in both Chinese and Indian defense growth, and India's greater volatility in defense spending levels, has two principal implications.

First, it will be important to monitor Indian and Chinese macroeconomic performance, since higher rates of GDP growth will make higher levels of defense and defense procurement spending more affordable by shrinking the defense share of GDP, holding other things constant. Second, it will be crucial to monitor Chinese and Indian policymakers' statements on the defense burden, i.e., the desired level of defense spending as a fraction of GDP, to ascertain their willingness to sustain defense spending beyond some nominal share of GDP, or to begin placing brakes on the growth in defense accounts. Indeed, the rather dramatic reduction in defense spending growth by both India and China in 2010 suggests that "guns versus butter" considerations already may be increasingly salient in each country's defense spending decisions.

Key Drivers of Defense and Defense Procurement Spending Estimates

As described above, our estimates of defense and defense procurement spending were sensitive to assumptions about what to include (e.g., defense share of Chinese defense industry group revenues) and exclude (e.g., Indian Ministry of Home Affairs paramilitaries), which data were available or missing (e.g., estimates of Chinese and Indian defense spending on defense nuclear and missile capabilities), assumed future average real GDP, defense spending, defense procurement growth rates over the 2010–2025 period, assumed future defense shares of GDP, and choice of exchange rate (market or PPP-based) for converting local currency units to constant U.S. dollars. That the resulting forecasts varied significantly depending on the specific combinations of these assumptions suggests the nature of the challenges of making predictions 15 years into the future. Nonetheless, China's starting advantages in GDP, defense and defense procurement spending, and a historical record of higher real growth rates in these aggregates confer a substantial edge over India that seems most likely to continue out to 2025.

Apples and Oranges: "National Defense" Versus "National Security" Spending

A key factor in comparing estimates of defense and defense procurement spending is navigating definitional differences regarding what is included in "national defense." Indeed, an important distinction between India and China is the way they each define national defense.

As described above, China includes its paramilitary force, the People's Armed Police, in its definition of national defense, even though spending on the PAP is not included in China's official defense budget estimate but rather is provided under China's Ministry for Public Security. By contrast, India's paramilitary forces, most of which are housed in the Ministry of Home Affairs, are not considered to be part of India's *national defense*, although they might be considered to be part of Indian *national security*.

To facilitate direct comparison between India and China, one could either include India's spending on paramilitaries (about 230 billion INR in 2009) or drop China's (75 billion RMB): The former approach would add about 14 percent to India's baseline estimate of 1,667 billion INR in defense spending in 2009–2010, whereas the latter approach would reduce our middle estimate of Chinese defense spending by about 10 percent. Thus, symmetric treatment of paramilitaries could reduce China's margin by 10–14 percent. Although this would not dramatically affect the basic thrust of our conclusions, these adjustments would make the bilateral comparisons more symmetric.

Spending on Strategic Nuclear, Missile, and Space Forces

It is not clear how much spending on the personnel, operations, and maintenance of the Chinese 2nd Artillery Corps, the component of the People's Liberation Army that controls China's nuclear ballistic and conventional missiles, is included in China's official estimate of national defense spending and how much might be hidden elsewhere in the Chinese defense or other (e.g., science and technology) budgets;[35] accordingly, the magnitude of this spending is largely unknown.

In a similar vein, spending on Indian strategic nuclear and missile forces was not broken out in the budget documents we reviewed. Moreover, most analysts believe

---

[35] Estimates of the 2nd Artillery Corps' share of Chinese defense spending appear to vary widely. For example, the Nuclear Threat Initiative estimates that "Proportionally, the Second Artillery Corps is given priority funding. Although it only makes up about 4 percent of the PLA, it receives 12 to 15 percent of the defense budget and about 20 percent of the total procurement budget." See Nuclear Threat Initiative (undated). GlobalSecurity.org argues that the 2nd Artillery Corps accounts for about half this percentage of the Chinese defense budget, or about 7 percent. See GlobalSecurity.org (2010).

Finally, although the credibility of the article is open to question, according to Wikipedia: "China categorizes 'the budget of the 2nd Artillery Corps' as 'the budget of Space Development Rockets,'" and "Missile development is included in the Air Science Budget." See Wikipedia (2010).

that research, development, and acquisition of Indian nuclear and missile forces take place within the Departments of Atomic Energy and Space, respectively, although the defense activities of these departments are not separately reported.

Thus, spending on defense nuclear and space activities remains a key area of uncertainty for both China and India.

Chinese Opacity and Indian Transparency in Defense Spending

Notwithstanding China's claims of transparency in its defense budget,[36] perhaps the most significant contrast between India and China is the level of visibility into Indian national defense and national security spending and the relative opacity of Chinese spending.

Among the major items that most analysts suggest are excluded from the official Chinese estimate of national defense expenditures are the PAP, local militia, overseas weapons procurement, defense industry subsidies, defense-related defense industry group revenues in excess of the official subtotal for defense spending on equipment, Chinese arms exports, R&D, and defense nuclear and missile capabilities.[37] For example, research, development, and acquisition spending is believed to take place within the "Equipment" account of China's official national budget estimate; the defense industry groups; and a host of scientific, technical, and other organizations; and much of this spending appears to be undocumented in publicly available documents, at least those in English.[38] Although some of these items are reported elsewhere in the Chinese state budget (e.g., PAP), other areas of uncertainty warrant further investigation.

In the case of India, four of the defense-related Demands for Grants in India's Union Budget—Capital Outlays, Ordnance Factories, Research and Development, and Ministry of Defence (for the Coast Guard Organisation)—account for the bulk of Indian defense-related research, development, and acquisition spending, not including spending related to defense nuclear and missile programs. Our analysis of budget documents revealed no equivalents to the Chinese defense industry groups or evidence of subsidies or other off-budget spending on defense or defense procurement. However, we do not rule out the possibility that such spending or subsidies take place within the Indian defense system.

---

[36] See, for example, Jiao (2010).

[37] See IISS (2006, pp. 249–253; 2010a, especially pp. 391–393).

[38] Shaoguang Wang estimated that defense-related R&D constituted 15 percent of the general R&D funds and 35 percent of the S&T budget. Cited in IISS (2006, p. 253).

Conclusions

The aim of the analysis presented here is to provide a comparative assessment of Indian and Chinese defense and defense procurement spending. We have not considered many strategic and political factors—for example, increased frictions or a warming of relations between China and India, a Mumbai-style attack that leads to increased spending to meet terrorist and related threats, or other strategic "wild cards." We have also not considered domestic political, organizational, or other societal changes that could affect defense spending in one or both countries. Such a relevant and complementary analysis would be valuable, but it was beyond the scope and resources of the present effort.

Our analyses also suggest several areas that warrant more detailed investigation and analysis.

For China, the many areas of opacity and uncertainty regarding its defense and defense procurement spending have already been emphasized. To realize a complete understanding of China's defense spending will present a continuing challenge for the future. That said, our analysis of Chinese spending relied primarily on English-language open-source materials, with a small amount of open-source Chinese materials providing additional, useful detail. Access to additional Chinese-language sources might reduce some of the opacity in Chinese defense and defense procurement spending. Increased transparency might also occur through Track Two meetings (i.e., non-government institutional meetings) with Chinese experts. This could help to identify additional data sources for estimating Chinese defense spending and to illuminate such issues as the magnitude of the overlap between official Chinese estimates of "Equipment"-related spending and the defense shares of defense industry group revenues.[39] Nonetheless, although China claims that its defense budget is completely transparent,[40] the full Chinese defense budget remains a state secret and Chinese interlocutors openly state that transparency is a tool that strong powers use to compel weaker parties to reveal their weaknesses and that China, being in a militarily inferior position to the United States, in fact has little incentive to improve transparency.[41]

In addition, it is also quite possible that some of India's defense and defense procurement spending is hidden from view, buried in other accounts, or provided for in unregistered inter-ministerial transfers. In particular, although Indian budget documents suggest that spending on defense-related nuclear and space activities is fairly modest, we have access only to aggregate data and cannot rule out the possibility that some spending is hidden. Also, there is little doubt that more detailed and exhaustive

---

[39] Indeed, there have been some notable recent efforts to improve the transparency of the Chinese defense budget through research and experts meetings. See, for example, Blasko et al. (undated).

[40] See, for example, Jiao (2010).

[41] See, for example, Yin (2009).

analysis of Indian defense and defense procurement is possible: For example, we did not have access either to the Budget Service Estimates that reportedly provided additional detail on India's armed services' spending or to the longer-range planning and budgetary documents, such as the Five Year Defence Plan, which is classified or at least sensitive. That said, we are unaware of other efforts to provide a comprehensive picture of Indian defense and defense procurement or to forecast spending through 2025. We hope that the present effort makes a contribution to these issues.

Furthermore, the volatility in recent Chinese and Indian defense spending—including greatly reduced growth rates for defense in 2010, as discussed above—reveals the inherent difficulties of using recent historical trends as a basis for forecasting future defense and defense procurement spending and the necessity of modesty in presenting any such forecasts. Although the present research has revealed some opportunities to improve our estimates of Chinese and Indian defense spending using unclassified sources and data, it may be that other approaches—employment of decisionmaking models or arms race models, for example—also could help to illuminate some of the factors that go into defense spending decisions.[42]

Finally, an important technical issue that was beyond the scope of the present effort is how to construct a "market basket" of defense goods to provide a better basis for PPP-based conversion of local currency units to U.S. dollars. This is an important methodological issue for consideration in future research.

In closing, we stress the need for additional empirical research to further illuminate and bound the many uncertainties associated with estimating and forecasting Chinese and Indian defense and defense procurement spending. We also note a need to better understand and model the factors that are likely to drive future Chinese and Indian defense spending decisions, including the many strategic and domestic "wild cards" that could affect the future trajectory of Indian and Chinese defense and defense procurement spending.

---

[42] One reviewer of this report suggested the potential utility of using Richardson arms race models to better understand the dynamics of Chinese-Indian defense spending. Use of agent-based rational choice stakeholder or expected utility models also could be a fruitful approach.

CHAPTER SIX

Conclusions and Implications

Two centuries ago, Edmund Burke advised a British parliamentarian: "Never plan the future by the past."[1] His admonition recalls a more recent truism, often attributed to Yogi Berra: "It's dangerous to make predictions, especially about the future!"

Unfortunately, the four-dimensional assessment summarized in this report violates both of these precepts. We have used data from the past and the present to make forecasts about the future. Consequently, we should reiterate the caveats mentioned earlier about the uncertainties surrounding our forecasts.

The enormous uncertainty involved in a comparative assessment of the status and performance of China and India 15 years from now can, in part, be reflected by envisioning multiple scenarios on how past conditions might change drastically in the future. To be sure, the scenarios—high-low and low-high pairings, optimistic and pessimistic scenarios—are themselves based on data and trends from the past. Numerous factors—political, economic, military, and both the domestic and international factors mentioned earlier—have been largely ignored in assessing the four domains, and these factors could substantially alter the forecasts for China and India in each of these domains (perhaps less so for the demographic forecasts than for those of the other three domains). Nevertheless, we have paired the respective high estimates for one country with low estimates for the other, and conversely, in order to highlight the uncertainties that might differentially impact the relative position of the two countries through 2025.

With these cautionary thoughts in mind, our concluding comments return to the key questions referred to at the outset: namely, in comparing India and China in 2025, Who will be ahead? Why? and By how much? More qualitatively, what are the advantages and disadvantages that each country will have in the four domains of our assessment?

---

[1] It is perhaps also relevant to recall George Santayana's countervailing advice: "Those who cannot remember the past are condemned to repeat it."

Demography

India's current rate of annual population growth is about twice that of China (1.55 percent versus 0.66 percent); as a consequence, its population will equal China's in 2025 (at about 1.4 billion) and exceed it thereafter. India's population is projected to continue increasing at least through 2050, while China's will reach its maximum (about 1.5 billion) in 2032, declining thereafter.

From the standpoint of effects on future economic performance, the changing age distributions of each country's population are more significant than their respective sizes. A larger percentage of China's population has been of working age (between 15 and 64), compared with India's, for the past three decades. The percentage of China's population that is of working age will peak by 2012, declining sharply thereafter. By contrast, the proportion of India's population that is of working age will continue to rise into the early 2030s, overtaking that of China in 2028. Hence, the potential labor productivity of India's total population will be increasing through the early 2030s and will surpass that of China by 2031. Whether and to what extent this potential Indian advantage is realized will depend on the employment opportunities available to India's rising cohorts of young workers, especially those for new entrants into the labor market.

Reflecting the changing age composition of their populations, China and India will experience quite different changes in their respective dependency ratios, i.e., the ratio of the over-65 and under-15 age cohorts to the prime working-age population. China's overall dependency ratio is currently nearly 39 percent. It will rise substantially in the next two decades, as the number of elderly dependents increases more rapidly than the number of child-age dependents decreases (the latter reflecting the effects of China's continued one-child policy).

The pattern of India's dependency ratios is quite different from China's. Currently, India's aggregate dependency ratio is 48 percent, 15 percent higher than China's. In India, the elderly make up a lower proportion of the overall number of dependents than in China, and this proportion is rising much more slowly in India than in China. The proportion of India's dependents that are young will increase slowly over the next decade and then increase moderately thereafter.

For the next two decades, India's dependency ratio will remain above China's, although the margin between their dependency ratios will decrease, as China's ratio will be rising and India's will be falling. After 2031, India will experience the advantage of a lower dependency ratio, while China will face a sharply rising one. Moreover, through 2025 and beyond, the changes in India's dependency ratios, as with every demographic indicator, are projected to be gradual, whereas changes in China will be more rapid and therefore more socially stressful.

Numerous other factors will affect the balance of advantages and disadvantages resulting from these demographic conditions. These factors include health, education, gender ratios, infrastructure, and migration. For example, China's population is generally healthier than India's, has the benefit of a more developed health care system, and has more experience in containing the effects of communicable diseases. On the other hand, China also has the impending burden imposed by a rising number and proportion of older people, so China's health costs can be expected to rise relative to those of a lesser-burdened India. India's younger population will be less prone to the costs of communicable and noncommunicable diseases.

China's population has, on average, a higher level of literacy and is better educated than India's. The smaller cohorts entering China's labor force in the 2020–2025 period will be more educated than the larger ones in India, providing an advantage for China and a challenge for India. To the extent that India successfully meets this challenge by effectively investing in human capital, it may achieve an advantage through productive employment of its growing pool of younger workers. Furthermore, the enhancement of human capital can be propelled by the allocation and effective use of research and development in government, business, and higher education, as discussed in Chapter Four and later in this chapter under "Science and Technology."

Finally, while the data on gender ratios in both China and India are unreliable as well as imperfectly comparable, both countries evidently have large gender imbalances, ranging between 17 percent and 30 percent more males than females in younger age cohorts. In China, the large imbalance is due to the combined effects of its one-child policy as well as gender preference, while India's equally large imbalance is due to the single effect of gender preference. How each of the two countries will manage the potential societal pressures arising from gender imbalance is both important as well as unknown.

Whether India's several demographic advantages—increasing numbers, younger age cohorts, declining dependency ratios, smoother transitional trends, etc.—will add up to a dividend or a drag on future growth depends on the extent to which productive employment opportunities emerge from an open, competitive, innovative Indian economy. A favorable answer to the question implies a substantial demographic dividend; an unfavorable one would lead to a demographic drag resulting from high unemployment, growing welfare burdens, and perhaps political unrest.

Conversely, whether China's several demographic disadvantages—rapid population aging, peaking and then declining population size, rising dependency ratios, increased gender imbalances, etc.—will create a drag or a dividend depends on whether these demographic circumstances trigger compensatory stimuli to improve technology, increase investment in human capital, or develop a more skilled and more productive labor force. If the compensatory effects are sufficient, China's demographic adversities may result in no drag on, and obliquely provide a dividend for, China's sustained economic growth.

Macroeconomics

In a recent speech in Washington, D.C., India's prime minister, Manmohan Singh, observed,

> There is no doubt that the Chinese growth performance is superior to Indian performance. . . . But . . . there are other values which are more important than the growth of the gross domestic product . . . the respect for fundamental human rights, the respect for the rule of law, respect for multicultural, multi-ethnic, and multi-religious rights. (Singh, 2009)

Notwithstanding the prime minister's delineation of relative importance, the focus of our macroeconomic assessment is GDP and its growth: past, present, and prospective. Chapter Three occasionally touched on the "more important" values mentioned by Singh, such as the rule of law, but the assessment has largely neglected them. Moreover, our assessment of the economic growth performance of India and China through the 2020–2025 period is moderately more favorable toward India than the prime minister's valid assessment of their respective performance up to the present.

The meta-analysis described in Chapter Three focused on forecasted growth of GDP, capital, employed labor, and total factor productivity in the two countries through 2025. What is striking about these results is the remarkably narrow margins between all of the paired China-India comparisons. The forecasted average GDP annual growth rates for 2020–2025 are 5.7 percent for China and 5.6 percent for India. The corresponding maximum GDP growth rates are 9.0 percent and 8.4 percent, respectively, and their paired minimum growth rates are 3.8 percent and 2.8 percent.

While our estimates for the Indian and Chinese growth rates are surprisingly close to one another, the absolute size of the GDP gap in China's favor would increase by 2025 because of the three-to-one greater size of China's initial GDP in 2010. Using the mean growth rates from the meta-analysis results in GDP estimates of $6.5 trillion for China and of $2.1 trillion for India in 2025, in constant 2000 prices at market exchange rates.

In a further effort to reflect as well as to bound the uncertainties of the meta-analysis forecasts, our assessment shows the GDP comparisons between India and China that result from five differing paired scenarios of their respective high, low, and average growth rates. As shown in Figures 3.5 and 3.6 of Chapter Three, only in the high-growth India/low-growth China scenario does India's GDP in 2025 approach China's. Using PPP conversion rates, in this scenario India's GDP reaches $12.3 trillion in 2025, while China's reaches $13.8 trillion. In the four other paired comparisons, China's GDP exceeds India's by factors that range between two and six!

In a further effort to glean information from the 27 studies included in the meta-analysis, we grouped them into three separate clusters reflecting their different organizational sponsorships: academic, business, and international agencies.

As discussed in Chapter Three, the business cluster is distinctly bullish (optimistic) about India's growth prospects and, relatively, bearish (pessimistic) about China's. The average annual growth rates projected in the business cluster for the 2020–2025 period for China and India are 4.7 percent, and 6.3 percent, respectively. By contrast, the two other clusters reverse this order. The academic cluster yields average growth rates of 5.5 percent for China and 4.3 percent for India, while the international-agencies cluster shows average growth rate estimates of 6.8 percent and 6.2 percent for China and India, respectively.

Chapter Three presents several conjectures (hypotheses) to explain the sharp differences among the clusters. For example, the business cluster's strong growth estimates for India relative to China perhaps derive from greater emphasis on a favorable business environment resulting from the rule of law, protection of property rights, and a democratic political system (this recalls the previously cited observations of Prime Minister Singh about other, "more important" values). Other hypotheses are presented in Chapter Three to account for the wider variance in estimates from the academic cluster studies and the more favorable estimates for China's growth from the business cluster studies.

Chapter Three concludes with our judgments about the India-China comparison pertaining to seven qualitative factors touched on in the 27 pooled studies: democracy and rule of law; IT and service skills, institutional stability; property rights; productivity growth; foreign investments in and by each country; and infrastructure. We conjecture that India has advantages in the first four of these, China in the next three. Whether this will enable India to catch up to or exceed China's "superior economic performance"—to use Prime Minister Singh's words—is arguable, as well as worthy of further study.

Science and Technology

In comparing the status and prospects of science and technology in China and India, we have focused on input indicators comprising both finance and human resources, as well as two output indicators. The financial inputs consist of gross domestic expenditures on R&D (GERD) as a percentage of GDP and its four components: R&D spending by higher education (HERD), by business (BERD), by government (GOVERD), and by private nonprofit institutions (PNPERD). The human resource inputs are doctoral degrees in science and engineering, including engineering, life sciences, physical sciences, computer science, mathematics, and agriculture.

The output measures used are publications (i.e., articles published in refereed scientific journals) and patents (especially triadic patents registered in the United States, the European Union, and Japan). These outputs are, at best, incomplete and imperfect. Many of the major effects of S&T inputs result in innovations, improvements in production and management processes, and more effective service delivery that are reflected in operational efficiency but not in either patents or publications.

While recognizing the limitations of these metrics, the assessment in Chapter Four uses them to compare India's and China's recent accomplishments and to develop a simulation model for projecting each country's future trajectory.

China is currently the world's third-largest R&D spender, with an annual growth rate of 18 percent since 2000. According to the OECD, China follows the United States and Japan in total R&D outlays (GERD). The business component (BERD), which may have the greatest early effects on productivity among the four components, has increased from 0.25 percent of China's GDP in 1996 to over 1 percent in 2006. Since then, China's GERD has increased in both absolute amounts and as a share of the country's GDP.

India's GERD represents 0.8 percent of its GDP, which is between one-third and one-half that of China (depending, respectively, on whether market exchange or PPP conversion rates are employed in comparing outlays in rupees and renminbi). India plans to raise its R&D to 2.5 percent of GDP in the next several years, perhaps stabilizing or slightly raising it still further through the 2025 period.

As described in Chapter Four, China currently graduates 70 percent more engineers than India annually (600,000 versus 350,000, respectively), although the two countries' data are imperfectly comparable. Apart from the reliability and comparability of these numbers, it is also difficult to assess the quality of ostensibly similarly credentialed engineers in the two countries. In a McKinsey Institute survey of 83 global multinational businesses (McKinsey Global Institute, 2005), the companies expressed their judgments of relative quality by saying they would hire only one in ten graduate engineers from China, but one in four from India. The substantial difference in inferred quality was ascribed to the Indian graduates' greater hands-on experience, as well as their better English-language and communication abilities.

To forecast the trajectories of China's and India's S&T growth, the assessment develops and applies a simulation model with the input variables mentioned above, along with parameters for costs and outputs (in credentialed S&T graduates, as well as publications and triadic patents per researcher). The model has been run with different values for these parameters, sometimes basing them on current levels prevailing in India and China and sometimes basing their values on current levels in South Korea on the plausible, but uncertain, assumption that the cost and productivity parameters realized by India and China during the next 15 years will converge toward those prevailing in South Korea in 2008. The several scenarios covered by the assessment include different combinations of these parameter values.

Chapter Four describes a wide range of results from these scenarios. Whether the bottom-line estimates for 2025 are represented by total numbers of full-time science and engineering researchers, holders of doctoral S&E degrees, or numbers of triadic patents or publications, the estimates for China exceed those for India by wide margins. For the simulation estimates of researchers and S&E publications in 2025, China's estimates exceed India's by factors of 8 and 13, respectively. In the scenarios in which we have adopted the McKinsey survey's 60 percent discount on quality (i.e., employability) of China's S&E doctorates to establish equivalence with India's, the margins decrease substantially: The numbers of doctoral degrees and their associated triadic patents are, for researchers and S&E publications, respectively, 1.5 and 1.7 times larger for China than for India.

Defense and Procurement Expenditures

As described in Chapter Five, our assessment of spending for defense and defense procurement by India and China employs the basic budget materials published by both countries in recent years to identify the detailed components as well as aggregate allocations over the past decade in each country. These materials are supplemented by other information sources to calculate the defense and defense procurement shares represented in the respective GDPs of China and India, how these shares have evolved, and the two countries' plans and prospects for maintaining or raising these shares.[2] To establish comparability between China and India, we have made adjustments to allow for differing definitions as to what each country includes within its defense spending estimates. For example, China includes paramilitary forces' expenditures in national defense spending, while India does not. Both countries exclude expenditures on nuclear weapons and delivery capabilities, although there is no presumption that the bilateral omissions are equal.

It is difficult to obtain reliable data on defense procurement from the official budgets of both countries, but the difficulties are much more acute for China. More generally, the gaps in China's official data coverage are larger and more opaque than those for India. In any event, the sequential approach that our assessment employs builds upon the official data for both countries to enable us to make estimates of each country's total expenditures on defense and on defense procurement, and to express these as shares of their respective GDPs.

Forecasts of these expenditures through 2025 are made using two different methods. The first method is based on a continuation of recent year-over-year real growth rates of defense spending; the second is based on defense spending as a fixed share of GDP, and linked to rising GDPs drawn from Chapter Three. The first method generates higher forecasts than the second; indeed, the estimates of the first method prob-

---

[2] There are frequently large differences among these "other information sources" in their estimates of China's defense spending. As indicated in Chapter Five, these estimates for 2003 differ by an order of magnitude between the official and highest unofficial estimates. The method we follow starts with the official budget figures, and then proceeds systematically to add elements we believe to be related to defense and to defense procurement, but which are lodged in other non-defense budget categories.

ably would be politically and/or financially infeasible because of the excessively high budget and GDP shares they would represent. Each of the two estimating methods was used to generate three alternative results, depending on whether optimistic, pessimistic, or moderate ("best") assumptions are made.

According to the first method, our "best" estimate for India's defense spending in 2025 is between $94 billion and $277 billion in constant 2009 dollars, depending on whether market exchange rates or PPP conversion rates are used. The corresponding "best" estimates for China's defense spending in 2025 lie between $688 billion and $1,258 billion. Our estimates of China's defense spending in 2025 are between four times and seven times those of India.

As noted, the forecasts generated by the GDP-based, second method are appreciably lower than those of the first method. For India, our "best" estimates of defense spending in 2025 lie between $82 billion and $242 billion in constant (2009) dollars; those for China are between $267 billion and $488. According to this method, China's defense spending would be between two and three times that of India in 2025.

Turning to defense procurement, we employ a method analogous to the first method referred to above for estimating total defense spending: namely, initially deriving for both China and India a 12.8 percent annual real growth of procurement spending from the data for the decade preceding 2009 levels. On this basis, our "best" estimates of defense procurement for India in 2025 are between $63 billion and $186 billion in constant dollars, depending respectively on whether market exchange rates or PPP rates are used to convert from rupees to dollars. The corresponding estimates for China's defense procurement spending in 2025 lie between $266 billion and $486 billion in constant dollars. Procurement spending in China would, by 2025, be about between 2.6 times and four times that of India.

Implications

It is worth repeating our early cautionary remarks about the abundant uncertainties surrounding these estimates. While there may be an excess of skepticism in the observation of Nobel Laureate Paul Samuelson that "Wall Street indexes predicted nine of the last five recessions,[3] his ironic remark is nonetheless a useful reminder lest our forecasts be taken too literally. Circumstances in India and in China may make our forecasts for defense spending and defense procurement very different from what actually ensues by 2025. Similar caution is warranted concerning our other forecasts for the two countries' demographics, economic growth, and scientific and technological development.

---

[3]   Samuelson (1966).

Although such forecasts are surrounded by uncertainties that could have large effects on any or all of the four domains we have investigated, this does not imply that these effects will change the positions of India and China *relative to one another.*

With this in mind, and while acknowledging the uncertainties and the ample grounds for caution in drawing conclusions, what can be said about the salient questions raised at the outset about China and India's relative advantages and disadvantages? And can we answer the three questions, Who's ahead? By how much? and Why?

The demographic assessment indicates several distinct advantages for India. Its population will continue to increase in size through 2025; the share of its population that is of prime working age is growing rapidly and will continue to do so beyond 2025. Its currently high dependency ratio is decreasing rapidly, and this will continue beyond 2025. In contrast, China's population will grow at a slow and decreasing rate, peaking several years after 2025 and declining thereafter. Its dependency ratios will be rising in the 2010–2025 period, and the rising costs of health care for the elderly will become an increasing burden. Gender imbalance is present in both countries but more severe in China, constituting a further source of demographic stress.

In sum, demographic changes are likely to be relatively more favorable to India than to China. From a developmental standpoint, demographic changes will provide a dividend for India and be a drag on the progress of China.

The three other dimensions of our assessment reverse this balance.

While the macroeconomic analysis in Chapter Three indicates that the average annual growth rates of India and China may be about equal over the next 15 years, the absolute difference between their respective GDPs is likely to increase in China's favor, simply because of the differences in starting points: China's current GDP is about three times larger than India's.[4] Whereas China's GDP in 2007 was $1.4 trillion larger than India's, in 2025, the difference between their respective GDPs will jump to $4.4 trillion, assuming both economies grow at the average annual rates shown in Chapter Three. So, our macroeconomic comparisons are relatively favorable to China.

A similar pattern emerges in the assessments of science and technology and of spending on defense and defense procurement, and for similar reasons. As with the macroeconomic assessment, the substantially larger base that China starts from generally results in higher absolute numbers for S&T outputs and for defense spending and procurement through 2025. Thus, our assessments for these two domains show distinct advantages for China.

The uncertainties accompanying these assessments have been reflected in various alternative scenarios—high and low, optimistic and pessimistic—that would significantly alter the paired comparisons. From the standpoint of drawing policy implications from our four-dimensioned comparisons between India and China, the several

---

[4]   Based on market exchange rates for the renminbi and rupee; using PPP rates makes China's GDP about two-and-a-half times larger than India's. See Figures 3.5 and 3.6 in Chapter Three.

paired scenarios in each domain can be regarded as building blocks that can be used by policymakers in China and India, as well as by policymakers in such third countries as the United States and Japan.

For example, in viewing the demographic prospects we have described, what are the national policy choices (including fiscal and monetary policies, as well as educational and health policies) that India can make to take full advantage of the potential demographic dividend we have envisaged? By the same token, what are the national policy choices that China can take that will avert or modulate the potential demographic drags we have described?

From the perspective of United States and other third countries, the alternative paired scenarios can serve as building blocks to identify and construct states of the paired China-India world in 2025 that appear more and less favorable to the United States or other third countries. For example, among the alternative S&T simulations, which scenarios appear most favorable to the United States, and can such scenarios be encouraged through foreign policy, alliance policy, or foreign trade and investment policies?

An important implication of such a building-block approach is that, by adopting suitable policies, the respective countries (i.e., India, China, and the United States as a prominent "third" country) may be able to affect the likelihood that one or another of the alternatives materializes, thereby altering the balance of advantages and disadvantages in the long-term, multifaceted competition between China and India.

Explicating the specific policies and their effects in altering our forecasted outcomes is worth further investigation, as well as beyond the purview of this study. That said, it is reasonable to suggest the following proposition: Prospects for India to pursue policies that will enhance its competitive position vis-à-vis China may be better than the opposite prospects for China. India's political-economic system entails at least a moderately greater degree of economic freedom compared with China's, and therefore India's environment may be more conducive to entrepreneurial, innovative, and inventive activity.[5]

---

[5]   In scaling "economic freedom," which is at most a suggestive rather than definitive indicator, India scores a few points better than China. See Heritage Foundation, 2010.

Meta-Analysis of Economic Growth in China and India

The first step in the meta-analysis involved collecting pertinent and accessible studies done between 2000 and 2008 addressing economic growth in China and India through the 2020–2025 period. The major sources consulted for the search included publication indexes, LexisNexis, and Internet search engines. This search yielded an initial pool of 47 studies.

The second step identified a subset of 27 studies that had the requisite data from India and China to permit their comparative assessment for the 2020–2025 period. Twenty studies were excluded because of incomplete or otherwise insufficient data for the two-country comparison.

The third step required drawing data from each study to make calculations of the recent and out-year rates of growth of GDP, employment, capital, and total factor productivity, either directly from the study or indirectly using incremental capital/output ratios, or a Cobb-Douglas production function, or a growth-accounting methodology. Twenty-seven of the 32 studies met these criteria for inclusion in the meta-analysis.

In the fourth step, these 27 studies and the corresponding descriptive statistics on GDP and factor growth rates, means, minima, maxima, and variances were arrayed into three separate groups or clusters of studies as follows:

- academic authors and institutions
- business organizations (e.g., Goldman Sachs, PricewaterhouseCoopers, McKinsey)
- international financial institutions (e.g., the World Bank, the International Monetary Fund, the Asian Development Bank).

This step included comparisons of the descriptive statistics across the three clusters of studies, and highlighting their similarities and differences.

In addition to the descriptive statistics, and the comparisons across the three clusters, the text discussion of the meta-analysis includes more detailed discussion of 17 of the 27 papers included in the meta-analysis.

Figure A.1 summarizes the successive steps.

Figure A.1
Meta-Analysis Process

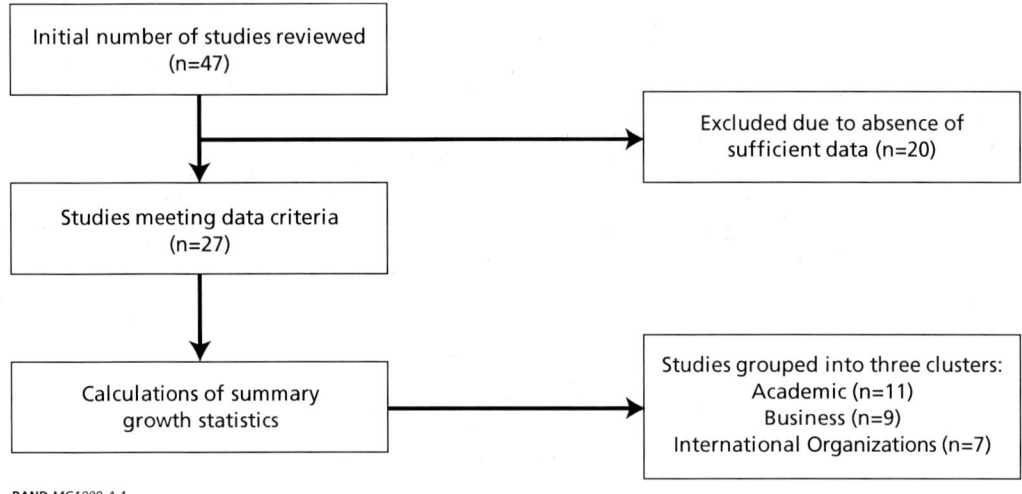

Detailed Calculations for, and Additional Figures Showing, the Projections in Chapter Four

1. **GDP:**

   Source of GDP: Chapter Three.

   Base year for GDP is 2008. Subsequent years are obtained by multiplying base years times the annual rate of growth. Mathematically, for instance, GDP in 2010 is calculated as follows:

   $$GDP_{t+s} = GDP_t * (1 + rg)^s \qquad (1)$$

   $$GDP_{2010} = GDP_{2008} * (1 + rg)^2 \qquad (1a)$$

   where rg = mean rate of economic growth.

2. **GERD as a percentage of GDP.**

   Source of GERD as a percentage of GDP: taken from the projections of China's and India's 11th Plans.
   Source of GERD in 2008 dollars: results from multiplying GERD as a percentage of GDP times the GDP in 2008 dollars for each year. Mathematically,

   $$GERD_t = gerd_t * GDP_t \qquad (2)$$

   where $gerd_t$= GERD as a percentage of GDP for year t.

3. **GERD components: average of each component over the decade 1995– 2005.**

Source of GERD components as a percentage of GDP in 2010–2025: average of Japan and South Korea over the 1995–2005 period.

GERD components in 2008 dollars: results from multiplying GERD as a percentage of GDP times the GDP in 2008 dollars for each year. Mathematically, for example,

$$HERD_t = herd \ * GDP_t \tag{3}$$

where $HERD_t$ = average HERD of South Korea and Japan as a percentage of GDP over the 1995–2005 period.

The same reasoning is applied to BERD, GOVERD, and PNPERD.

4. **Cost per FTR and total number of researchers.**

Source: Elaboration based on OECD, NSTMIS,[1] and NSF S&E (2008) data.

We first need to estimate the cost per FTR, which we do by dividing the total GERD in 2008 PPP U.S. dollars by the number of FTRs[2]:

$$CostFTR_{2008} = \frac{GERD_{2008}}{FTR_{2003}}. \tag{4}$$

5. **Cost per FTR per year.**

a. **Projections under alternative scenarios:** Two scenarios are considered as follows:

Scenario 1: current cost per FTR
Scenario 2: assuming South Korea's cost per FTR.

China's cost per researcher is well behind the rest of Asia and the developed world. Thus, South Korea's cost per FTR implies a huge increase in costs and, consequently, a reduction in the stock of FTRs.

b. **Scenario 1—Optimistic Scenario.**

---

[1]   *NSTMIS* refers to India's National Science and Technology Management Information System. See Government of India, Department of Science and Technology (undated).

[2]   This estimate could be biased upward as long as the number of FTRs is the last year available, 2003.

Number of FTRs is estimated based on current cost per FTR in each country.
Based on the cost per FTR 2008 we estimate the number of researchers for year t by dividing the GERD in year t expressed in 2008 PPP U.S. dollars by the cost per FTR_{2008} expressed in the same currency.

Thus, the increase in GERD associated with the increase in GDP allows acquiring FTR according to current cost.

$$TotalNumberofFTR_t = \frac{GERD_{t+s}}{CostFTR_{2008}} \tag{5}$$

c. **Scenario—Pessimistic Scenario.**

Number of FTRs is estimated based on South Korea's current cost per FTR.
In this case we first consider the relationship between cost per FTR in China and India in the base year, 2008, according to the following formula:

$$CostFTR_{i,t(SouthKorea)} = \frac{CostFTR_{i,2008}}{CostFTR\,SouthKorea_{2008}} \tag{6}$$

where $CostFTR_{i,t\,(SouthKorea)}$ expresses the cost of FTR in country i expressed in terms of cost per FTR in South Korea.

Total number of FTRs in country i in year t are obtained as follows:

$$FTR_{i,t(SouthKorea)} = \frac{GERD_{i,t}}{CostFTR\,SouthKorea_{2008}}. \tag{7}$$

6. **Total number of Ph.D. diplomas in S&E.**

To project the number of Ph.D. diplomas in S&E, we consider the current relationship between Ph.D. diplomas and FTRs in China and India. These are estimated by dividing the number of Ph.D.'s by the total number of FTRs in the base year:

$$PhD\_per\_FTR_{i,2008} = \frac{PhD\,Diplomas_{i,2008}}{Number\,FTR_{i,2008}}. \tag{8}$$

a. **Total number of Ph.D. diplomas in S&E.**

The projected number of Ph.D. diplomas will be estimated by multiplying the number of projected FTRs in each scenario by the relationship between Ph.D. diplomas and FTR:

$$PhD_{i,t} = PhD\_per\_FTR_{i,2008} * FTR_{i,t} \tag{9}$$

where $PhD_{i,t}$ expresses the total number of Ph.D. diplomas in S&E in country i in year t.

7. **Total Number of Patents:**

Source: Elaboration based on OECD, NSTMIS, and NSF S&E (2008).

Projections

Two alternative scenarios are again considered:

Scenario 1: projects number of triadic patents based on the current relationship between FTRs and triadic patents. Given the relative low productivity of Chinese and Indian researchers in terms of inventions with respect to the industrialized world this represents a pessimistic scenario.

Scenario 2: in this scenario both countries will converge to South Korea's current productivity per researcher which is higher than both China and India's current productivity per FTR.

Estimation of Current "Researcher Productivity": Current productivity or triadic patent per FTR is determined by the following relationship:

$$TP\_per\_FTR_{i,2008} = \frac{TriadicPatents_{i,2008}}{Number\ FTR_{i,2008}}.$$ (10)

where $TP\_per\_FTR_{i,2008}$ is the number of triadic patents per FTR in country i in the base year, 2008.

a. **Scenario 1**—Pessimistic Scenario.

Holds observed productivity per FTR in China and India as the basis for projections.

The estimated number of triadic patents for each year t will be equal to the number of FTR times the current productivity:

$$TP_{i,t} = TP\_per\_FTR_{i,2008} * FTR_{i,t}.$$ (11)

b. **Scenario 2**—Optimistic Scenario.

In this case we assume that FTR's productivity will increase to the one observed in Korea in the base year, 2008.

Thus, the number of triadic patents in country i in year t is estimated as follows:

$$TP_{i,t} = TP\_per\_FTR_{South\ Korea,2008} * FTR_{i,t} .$$ (12)

8.   Total Number of Publications in S&E

To project the number of publications in S&E, we consider the current relationship between publications and FTR in China and India. These are estimated by dividing the number of publications by the total number of FTRs in the base year:

$$Pub\_per\_FTR_{i,2008} = \frac{Pub_{i,2008}}{FTR_{i,2008}} .$$ (13)

a. **Estimation** of current "Researcher Productivity."
b. **Projection** of total number of publications.

Projections are the results of multiplying the number of FTRs in country i in year t times the productivity observed in country i in the base year, 2008 as follows:

$$Pub\_per\_FTR_{i,t} = Pub\_per\_FTR_{i,2008} * FTR_{i,t} .$$ (14)

Projections for Researchers, S&T Ph.D.'s Graduated, Patents, and Publications

Figures B.1–B.4 show our estimates for four indicators of progress in S&T—researchers, S&T Ph.D.'s graduated, patents, and published journal articles—for China and India through 2025 under the four different scenarios described in Chapter Four. As explained in Chapter Four and in the above calculation steps,

- Scenario 1 considers the current S&T parameters in China and India (Current Cost FTR/Current Productivity).
- Scenario 2 assumes that cost per FTR will converge to the level in South Korea (South Korea's Cost FTR/Current Productivity).
- Scenario 3 assumes that researcher productivity (the numbers of triadic patents and S&E publications per FTR) for China and India will converge to the level in South Korea (South Korea's Productivity FTR/Current Cost per FTR).
- Scenario 4 assumes an imputed productivity for Indian engineers 60 percent higher than that for Chinese engineers (based on McKinsey Global Institute [2005], China's factor is 0.12 and India's is 0.30).

Note that in Figures B.1–B.4 the y-axis labels for China and India are different. In each figure, the numbers for China are much larger than those for India, even though the heights of the bars in the charts may be similar.

Figure B.1
Researchers, India and China, 2008–2025 (thousands)

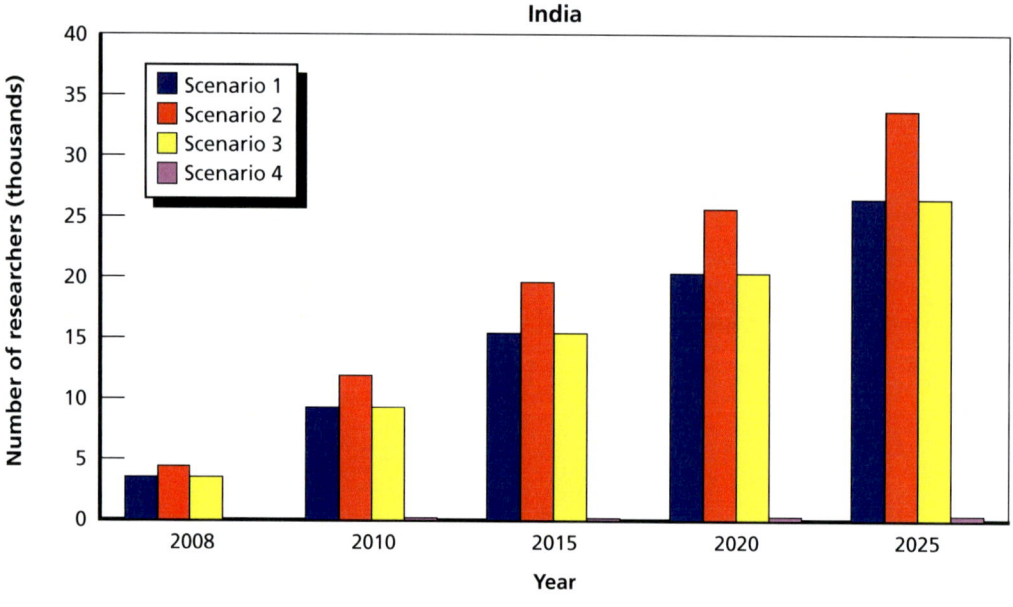

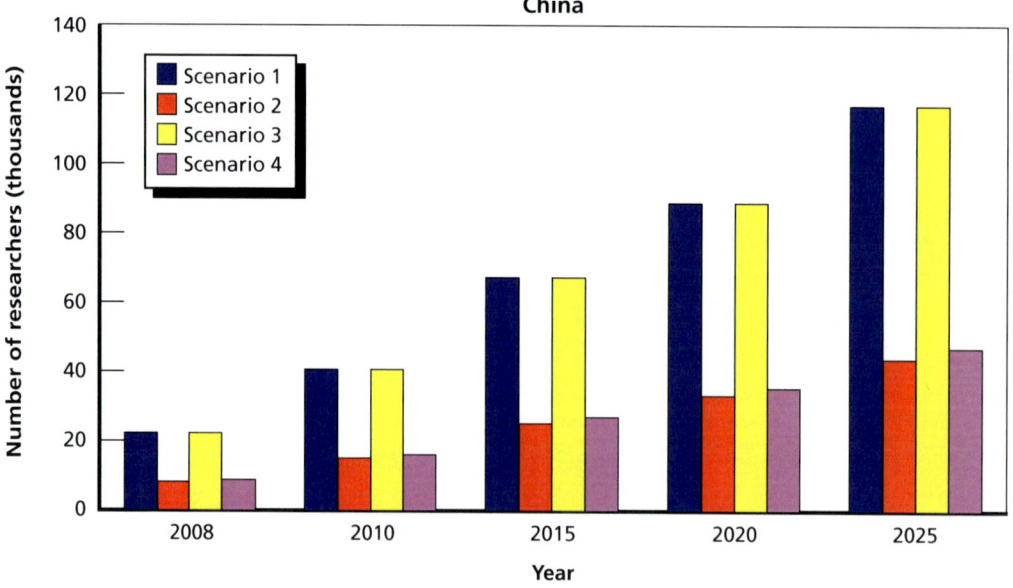

NOTE: The scales in the charts are different.

Figure B.2
Ph.D. Diplomas in S&T, India and China, 2008–2025 (thousands)

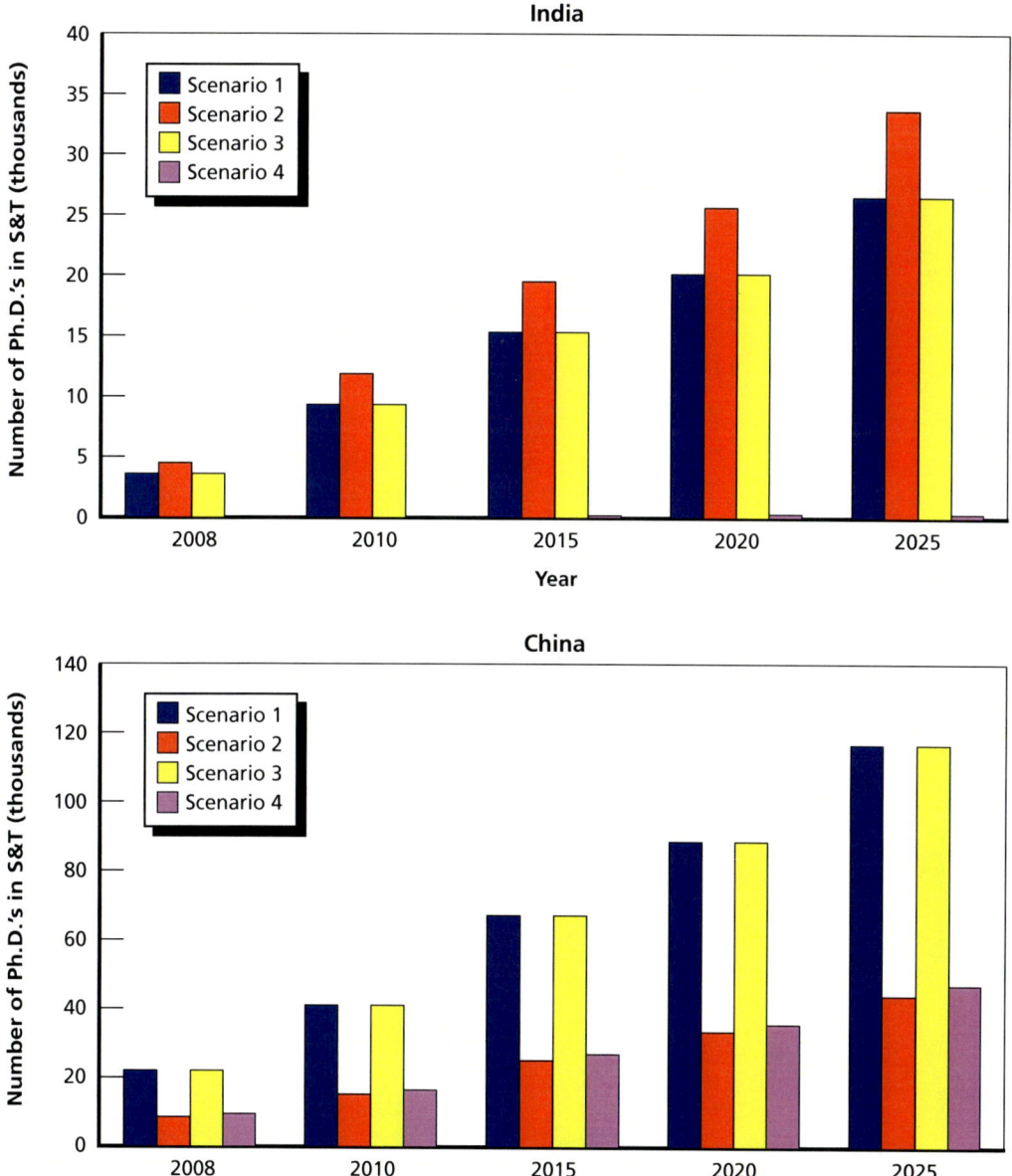

NOTE: The scales in the charts are different.

RAND *MG1009-B.2*

Figure B.3
Patents, India and China, 2008–2025

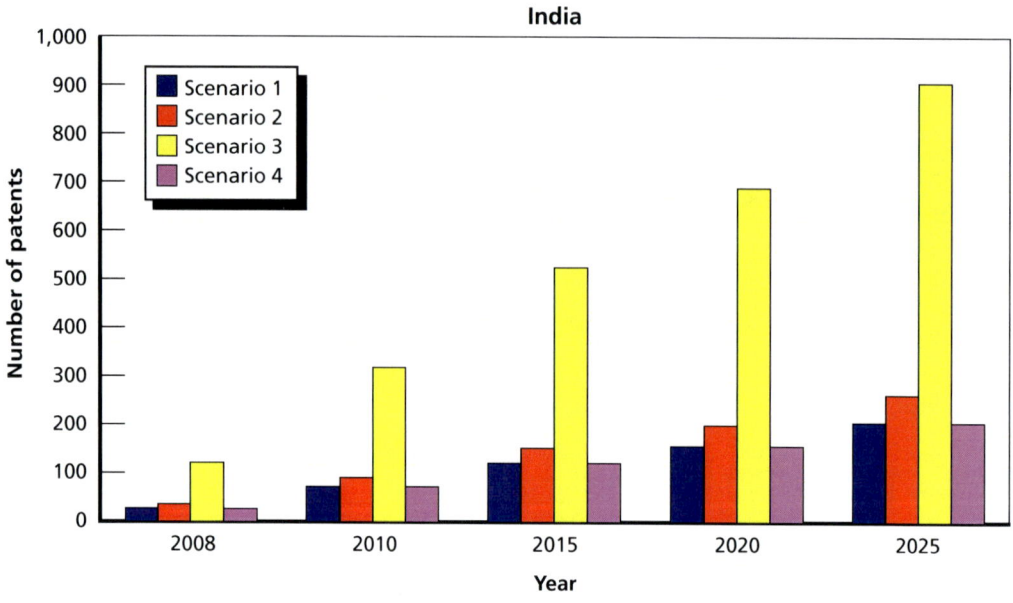

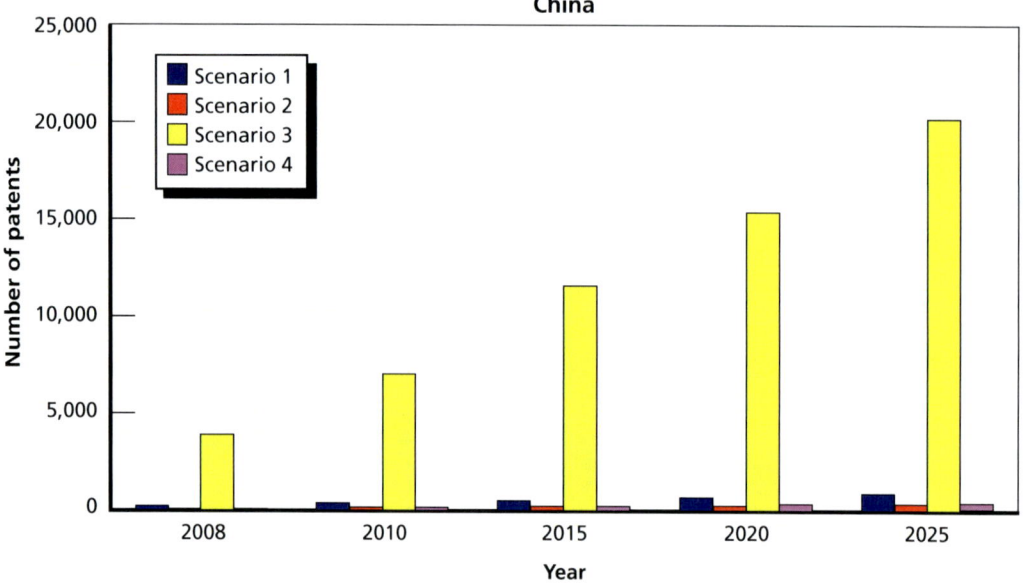

NOTE: The scales in the charts are different.

Figure B.4
Publications, India and China, 2008–2025 (thousands)

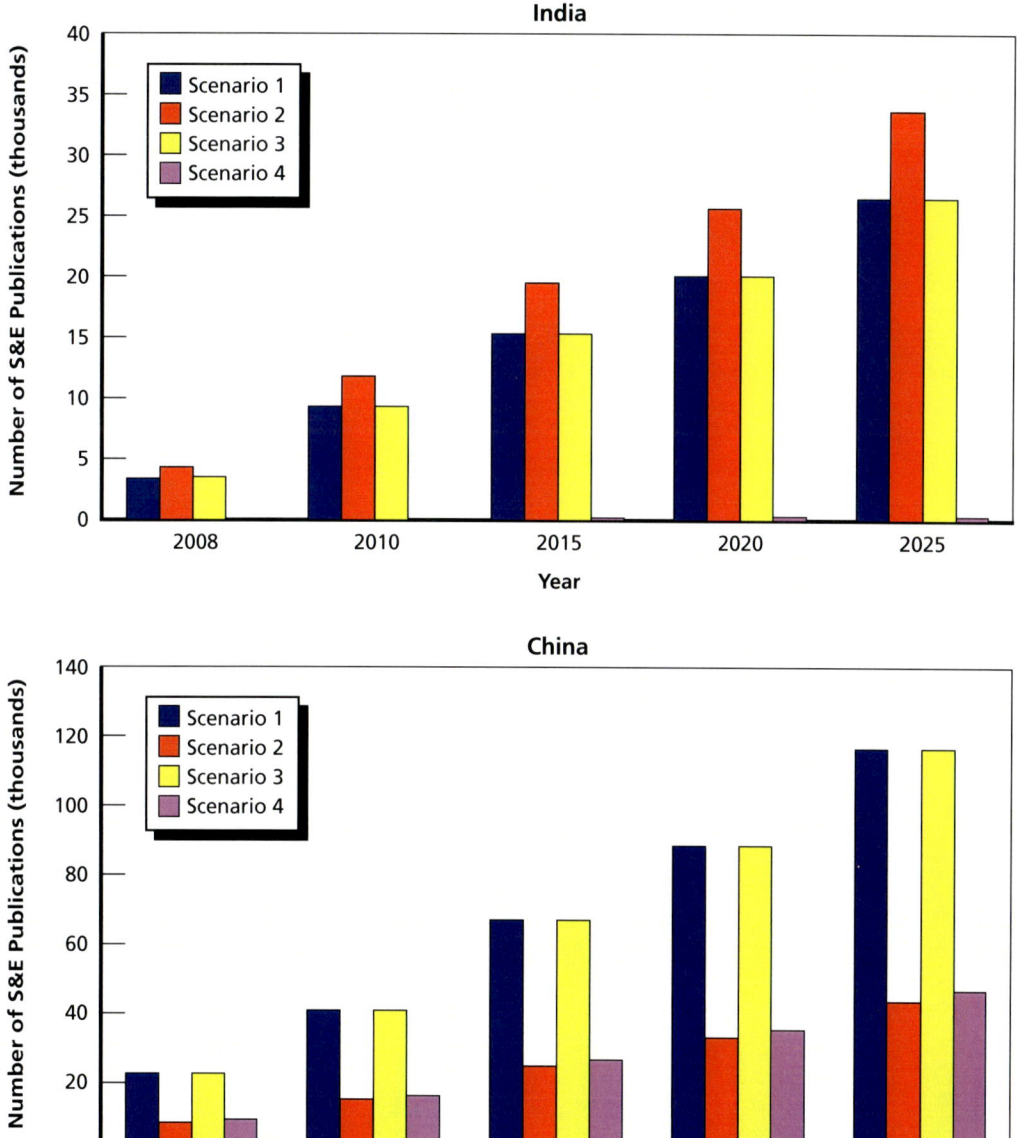

NOTE: The scales in the charts are different.

Analytic Tables

This appendix provides additional detail on the various assumptions behind our analyses and forecasts:

- Table C.1 summarizes our estimates of 2009 Chinese and Indian GDP levels, the average 2009 market and PPP-based exchange rates used to convert local currency units (LCUs, i.e., Chinese renminbi or Indian rupees) into U.S. dollars, and 2009 GDP estimates in U.S. dollars (MXR and PPP).
- Table C.2 presents data on nominal historical GDP levels, nominal year-on-year GDP growth, implicit GDP price deflators, and real year-on-year growth in GDP for China and India.
- Table C.3 presents data on nominal historical defense spending levels, nominal year-on-year growth, implicit GDP price deflators, and real year-on-year growth in defense spending for China and India.
- Table C.4 presents data on nominal historical defense procurement spending levels, nominal year-on-year growth, implicit GDP price deflators, and real year-on-year growth in defense procurement spending for China and India.
- Table C.5 presents the basis for our estimates of the total revenues and defense-related revenues of China's defense industry groups.

Table C.1
Summary of GDP, GDP Growth, Inflation and Exchange Rate Assumptions

	China	India
2009 GDP (billions of LCUs)	34,050.7	59,520.0
2009 LCU-USD exchange rate (MXR)	6.831416	48.405267
2009 LCU-USD exchange rate (PPP)	3.740117	16.468991
2009 GDP (billions of U.S. $, MXR)	4,984.4	1,229.6
2009 GDP (billions of U.S. $, PPP)	9,104.2	3,614.1

SOURCES: GDP estimates for 2009 are based on data from the International Monetary Fund (2010b). Exchange rates are from the World Bank (undated).

Table C.2
Gross Domestic Product: Nominal (Then-Year) and Real Levels and Year-on-Year Growth,
2000–2009

Year	China					India				
	Nominal RMB (B)	% Nominal Y-o-Y Growth	Implicit GDP Deflators 1999=100	Real 1999 RMB (B)	% Real Y-o-Y Growth	Nominal INR (B)	% Nominal Y-o-Y Growth	Implicit GDP Deflators 1999=100	Real 1999 INR (B)	% Real Y-o-Y Growth
1999	8967.7	–	100.0	8967.7	–	19678.3	–	100.0	19678.2	–
2000	9921.5	10.6	102.1	9720.9	8.4	21566.2	9.6	104.9	20552.1	4.4
2001	10965.5	10.5	104.2	10527.0	8.3	23189.1	7.5	108.6	21350.5	3.9
2002	12033.3	9.7	104.8	11485.1	9.1	24998.0	7.8	112.0	22323.5	4.6
2003	13582.3	12.9	107.4	12645.1	10.1	27732.7	10.9	116.3	23853.2	6.9
2004	15987.8	17.7	114.8	13921.2	10.1	31272.4	12.8	121.3	25786.7	8.1
2005	18493.7	15.7	119.4	15492.9	11.3	35708.6	14.2	126.8	28150.7	9.2
2006	21631.4	7.0	123.9	17459.3	12.7	41139.8	15.2	133.3	30869.2	9.7
2007	26581.0	22.9	133.3	19937.1	14.2	47633.5	15.8	140.4	33921.2	9.9
2008	31404.5	18.1	143.7	21850.0	9.6	54873.0	15.2	152.0	36090.5	6.4
2009	34050.7	8.4	142.8	23837.4	9.1	59520.0	8.5	156.0	38139.9	5.7
2000–2009	Average:	14.4		Average:	10.3		11.7		Average:	6.9
				CAGR:	10.3				CAGR:	6.8

SOURCE: GDP estimates and implicit GDP price deflators are from the International Monetary Fund
(2010b).
NOTES: Y-o-Y = year-on-year. CAGR = cumulative average growth rate.

Table C.3
Defense Spending: Nominal (Then-Year) and Real Levels and Year-on-Year Growth,
2000–2009

	China					India				
Year	Nominal RMB (B)	% Nominal Y-O-Y Growth	Implicit GDP Deflators 1999=100	Real 1999 RMB (B)	% Real Y-O-Y Growth	Nominal INR (B)	% Nominal Y-o-Y Growth	Implicit GDP Deflators 1999=100	Real 1999 INR (B)	% Real Y-O-Y Growth
1999	107.6	–	100.0	107.6	–	584.5	–	100.0	584.5	–
2000	120.8	12.3	102.1	118.4	10.0	602.7	3.1	104.9	574.4	–1.7
2001	144.2	19.4	104.2	138.4	17.0	650.8	8.0	108.6	599.2	4.3
2002	170.8	18.4	104.8	163.0	17.8	664.0	2.0	112.0	592.9	–1.0
2003	190.8	11.7	107.4	177.6	9.0	717.7	8.1	116.3	617.3	4.1
2004	220.0	15.3	114.8	191.6	7.8	897.1	25.0	121.3	739.7	19.8
2005	247.5	12.5	119.4	207.3	8.2	949.8	5.9	126.8	748.8	1.2
2006	297.9	20.4	123.9	240.4	16.0	1008.1	6.1	133.3	756.4	1.0
2007	355.5	19.3	133.3	266.6	10.9	1085.5	7.7	140.4	773.0	2.2
2008	418.3	17.7	143.7	291.0	9.1	1372.2	26.4	152.0	902.5	16.8
2009	480.7	14.9	142.8	336.5	15.6	1666.6	21.5	156.1	1067.9	18.3
2000–2009	Average:	16.2		Average:	12.1		11.4		Average:	6.5
				CAGR:	12.1				CAGR:	6.2

SOURCES: Historical data on nominal Chinese defense spending levels are from the State Council Information Office (undated, p. 103); National Bureau of Statistics of China (2008); and "China's Defense Budget to Grow 14.9% in 2009" (2009). Historical data on nominal Indian defense spending levels are from Government of India (1999–2000 through 2009–2010). Implicit GDP price deflators are from the International Monetary Fund (2010b).

NOTES: Nominal totals for China are official Chinese estimates of defense spending. Y-o-Y = year-on-year. CAGR = cumulative average growth rate.

Table C.4

Defense Procurement: Nominal (Then-Year) and Real Levels and Year-on-Year Growth, 2001–2009

	China					India				
Year	Nominal RMB (B)	% Nominal Y-o-Y Growth	Implicit GDP Deflators 1999=100	Real 1999 RMB (B)	% Real Y-o-Y Growth	Nominal INR (B)	% Nominal Y-o-Y Growth	Implicit GDP Deflators 1999=100	Real 1999 INR (B)	% Real Y-o-Y Growth
2000	38.9	–	102.1	38.1	–	126.3	–	104.9	120.4	–
2001	49.5	27.2	104.2	47.5	24.7	146.2	15.8	108.6	134.6	11.8
2002	57.3	15.8	104.8	54.7	15.1	125.8	–14.0	112.0	112.3	–16.5
2003	62.0	8.2	107.4	57.7	5.5	151.9	20.7	116.3	130.7	16.3
2004	74.8	20.6	114.8	65.1	12.8	285.3	87.8	121.3	235.3	80.1
2005	83.7	11.9	119.4	70.1	7.7	272.6	–4.5	126.8	214.9	–8.7
2006	97.3	16.2	123.9	78.5	12.0	349.3	28.1	133.3	262.1	22.0
2007	114.4	17.6	133.3	85.8	9.3	300.9	–13.9	140.4	214.3	–18.2
2008	139.1	21.6	143.7	96.8	12.8	323.2	7.4	152.0	212.6	–0.8
2009	160.0	15.0	142.8	112.0	15.7	429.7	33.0	156.1	275.3	29.5
2001–2009	Average:	17.1		Average:	12.8		17.8		Average:	12.8
				CAGR:	12.6				CAGR:	9.5

SOURCES: Historical data on nominal Chinese defense spending levels are from State Council Information Office (undated, p. 103); National Bureau of Statistics of China (2008); and "China's Defense Budget to Grow 14.9% in 2009" (2009). Historical data on nominal Indian defense spending levels are from Government of India (1999–2000 through 2009–2010). Implicit GDP price deflators are from the International Monetary Fund (2010b).

NOTES: Nominal totals for China are for the "Equipment" account of the official estimate. Y-o-Y = year-on-year. CAGR = cumulative average growth rate.

Table C.5
Spending Estimates for Defense Industry Groups, 2001–2008 (nominal RMB in billions)

	2001[a]	2002[a]	2003[a]	2004[a]	2005[a]	2006[bh]	2007[c]	2008[defgi]
CNNC	8.475	9.657	13.199	17.330	18.916	21.151	27.15	37.714
CNEC	(8.475)	(9.657)	(13.199)	(17.330)	(18.916)	(21.151)	(27.15)	(37.714)
CASC	(13.366)	(22.671)	(25.453)	(28.235)	(34.005)	(42.298)	50.59	(61.020)
CASIC	13.366	22.671	25.453	28.235	34.005	(42.298)	(50.59)	61.020
AVIC								151.075
AVIC I	24.501	36.111	(46.659)	58.058	69.989	82.092	104.81	
AVIC II	21.207	26.300	33.982	(37.546)	41.110	46.959	50.92	
CSSC	16.674	19.715	25.534	(45.804)	(66.074)	(86.344)	(84.021)	(81.697)
CSIC	17.712	20.048	29.139	41.937	49.641	64.451	82.10	103.483
NORINCO	35.571	42.615	52.118	64.061	79.411	105.959	133.88	147.584
CSG	25.085	39.692	50.159	64.355	75.222	(100.581)	(125.941)	151.3[f]
Memo: COEC					75.222	101.111	141.148	150.643
CETC	n/a	n/a	n/a	n/a	n/a	61.131	68.49	
GRAND TOTAL	(184.432)	249.137	314.895	402.891	487.289	674.414	805.641	908.932
Assumed defense share								
@20 percent ("low")	36.9	49.8	63.0	80.6	97.5	134.9	161.1	181.8
@25 percent ("best")	46.1	62.3	78.7	100.7	121.8	168.6	201.4	227.2
@30 percent ("high")	55.3	74.7	94.5	120.9	146.2	202.3	241.7	272.7
@35 percent	64.6	87.2	110.2	141.0	170.6	236.0	282.0	318.1

SOURCES: a. Surry (2007). b. "2007 Top 500 Chinese Enterprises List." c. "2008 Top 500 Chinese Enterprises List." d. "2009 Top 500 Chinese Enterprises List Ranking List (All)." e. "Top 500 Chinese Enterprises 2009 Released," (2009). f. "China South Industries Group Corporation Announces Earnings Results for the Year 2008" (2009). g. ResearchInChina (2009). h. "China Shipbuilding Industry Report, 2006–2007." i. CSSC revenue in 2008 is estimated at $15 billion, whereas CSIC revenue is estimated at $18 billion; CSSC is estimated as 15/19 of CSIC revenue of 103.483 billion yuan in 2008. See "China's Corporation in Technology Intensive Industries!" 2009.

NOTES: CNNC = China National Nuclear Corporation; CNEC = China Nuclear Engineering and Construction Corporation; CASC = China Aerospace Science and Technology Corporation; CASIC = China Aerospace Industry Corporation; AVIC = China Aviation Industry Corporation; CSSC = China State Shipbuilding Corporation; CSIC = China Shipbuilding Industry Corporation; NORINCO = China North Industries Group Corporation; CSG = China South Industries Group Corporation; CETC = China Electronics and Technology Corporation; COEC = China Ordnance Equipment Group Corporation, a subsidiary of the China South Industries Group. Numbers in parentheses are authors' estimates. Total estimated shipbuilding revenue in 2006 is estimated at 150.795 billion yuan; subtracting 64.451 billion yuan for CSIC yields 86.344 billion yuan for CSSC.

References

"2007 Top 500 Chinese Enterprises List." As of November 2009:
http://www.cec-ceda.org.cn/huodong/2007china500/11.htm

"2008 Top 500 Chinese Enterprises List." As of November 2009:
http://finance.sina.com.cn/hy/20080830/10375254697.shtml

"2009 Top 500 Chinese Enterprises List Ranking List (All)." As of November 2009:
http://www.ce.cn

Ablett, Jonathan, et al., *The 'Bird of Gold': The Rise of India's Consumer Market*, McKinsey Global
Institute, 2007.

Acharya, Shankar, "Can India Grow Without Bharat?" *India Abroad*, November 25, 2003. As of July
20, 2010:
http://www.rediff.com/money/2003/nov/25guest.htm

Acs, Zoltan J., David B. Audretsch, Pontus Braunerhjelm, and Bo Carlsson, "Growth and
Entrepreneurship: An Empirical Assessment," Max Planck Institute of Economics, Papers on
Entrepreneurship, Growth and Public Policy 2005-32, 2005.

Aghion, Philippe, and Peter Howitt, "A Model of Growth Through Creative Destruction,"
Econometrica, Vol. 60, 1992, pp. 323–351.

Ahya, Chetan, Andy Xie, Stephen S. Roach, Sheth Mihir, and Denise Yam, *India and China: New
Tigers of Asia, Part II*, Morgan Stanley Special Economic Analysis, June 2006. As of July 21, 2010:
http://www.scribd.com/doc/7291150/Comparison-Between-India-and-China

Akteruzzaman, M. D., "China's Economic Development and Export Promotion Strategy: Can
Bangladesh Learn?" *The Social Sciences*, Vol. 1, No. 6, 2006.

Anderson, Jamie, and Costas Markides, "Strategic Innovation at the Base of the Pyramid," *MIT
Sloan Management Review*, Vol. 49. No. 2, Fall 2007. As of July 21, 2010:
http://sloanreview.mit.edu/x/49116

Apps, Patricia Frances, and Ray Rees, "Fertility, Female Labor Supply and Public Policy," Institute
for the Study of Labor (IZA), Discussion Paper #409, 2001.

Asuncion-Mund, Jennifer, *India Rising: A Medium-Term Perspective*, Deutsche Bank Research, March
19, 2005.

Banister, Judith, David E. Bloom, and Larry Rosenberg, "Population Aging and Economic Growth
in China," Program on Global Demography and Aging, Harvard University, Working Paper 53,
March 2010. As of July 21, 2010:
http://www.hsph.harvard.edu/pgda/WorkingPapers/2010/PGDA_WP_53.pdf

Bardhan, Pranab, "Crouching Tiger, Lumbering Elephant: A China-India Comparison," in Kaushik Basu, Ranjan Ray, and Pulin Nayak, eds., *Markets and Governments*, New York: Oxford University Press, 2003.

Behera, Laxman K., "An Analysis of Defense Budget 2010–11," *New Delhi Political and Defence Weekly*, Vol. 9, No. 23, March 9–15, 2010, pp. 15–17.

Bellman, Eric, "Indian Firms Shift Focus to the Poor," *Wall Street Journal*, October 21, 2009. As of July 21, 2010:
http://online.wsj.com/article/SB125598988906795035.html

Bergheim, Stefan, *Global Growth Centers 2020: "Formel-G" for 34 Economies*, Deutsche Bank Research, March 23, 2005.

Bergsten, C. Fred, Bates Gill, Nicholas R. Lardy, and Derek Mitchell, *China: The Balance Sheet: What the World Needs to Know Now About the Emerging Superpower*, Center for Strategic and International Studies, Cambridge, Mass.: Perseus Group, 2006.

Bernardes, Américo Tristão, Ricardo Machado Ruiz, Leonardo Costa Ribeiro, and Eduardo da Motta e Albuquerque, *Modeling Economic Growth Fuelled by Science and Technology*, Universidade Federal de Minas Gerais, Facultua de Ciencias Economicas, Centro de Desenolvivimento e Planeamiento Regional, 2006.

Bitzinger, Richard A., "China's Defense Budget: Is the PLA Cooking the Books?" *International Defense Review*, February 1995.

———, "Just the Facts, Ma'am: The Challenge of Analysing and Assessing Chinese Military Expenditures," *The China Quarterly*, No. 173, March 2003, pp. 164–175.

Bitzinger, Richard A., and Chong-Pin Lin, *The Defense Budget of the People's Republic of China*, Washington, D.C.: Defense Budget Project, November 1994.

Blasko, Dennis J., Chas W. Freeman, Jr., Stanley A. Horowitz, Evan S. Medeiros, and James C. Mulvenon, *Defense-Related Spending in China: A Preliminary Analysis and Comparison with American Equivalents*, United States–China Policy Foundation, undated. As of July 21, 2010:
http://www.uscpf.org/v2/pdf/defensereport.pdf

Bloom, David E., with Alexia Prskawetz and Wolfgang Lutz, eds., "Population Aging, Human Capital Accumulation, and Productivity Growth," a supplement to *Population and Development Review*, Vol. 33, 2007.

Bloom, David E., and David Canning, "Global Demographic Change: Dimensions and Economic Significance," in David E. Bloom, Alexia Prskawetz, and Wolfgang Lutz, eds., *Population Aging, Human Capital Accumulation, and Productivity Growth*, special issue of *Population and Development Review*, supplement to Vol. 33, 2007.

Bloom, David E., David Canning, Günther Fink, and Jocelyn E. Finlay, "Fertility, Female Labor Force Participation and the Multiplier Effect," Harvard School of Public Health, 2009. As of July 21, 2010:
http://iussp2009.princeton.edu/download.aspx?submissionId=92713

Bloom, David E., David Canning, and Jaypee Sevilla, *The Demographic Dividend: A New Perspective on the Economic Consequences of Population Change*, Santa Monica, Calif.: RAND Corporation, MR-1274-WFHF/DLPV/RF, 2003. As of July 21, 2010:
http://www.rand.org/pubs/monograph_reports/MR1274.html

Bosch, Mariano, Daniel Lederman, and William F. Maloney, "Patenting and Research and Development: A Global View," Washington, D.C.: World Bank, Policy Research Working Paper Series 3739, 2005.

Bosworth, Barry, and Susan M. Collins, "Accounting for Growth: Comparing China and India," *Journal of Economic Perspectives*, Vol. 22, No. 1, 2008, pp. 45–66.

Bradsher, Keith, "In Downturn, China Sees Path to Growth," *New York Times*, March 17, 2009.

Brown, Harold, "Managing Change: China and the United States in 2025," address at the 8th Annual RAND-China Reform Forum Conference, RAND Corporation, June 28, 2005, RAND Corporation, 2005. As of July 21, 2010:
http://www.rand.org/pubs/corporate_pubs/CP505.html

Cai, April, "China's Pension System Faces Major Reform," *China Business*, July 21, 2006. As of July 21, 2010:
http://www.atimes.com/atimes/China_Business/HG21Cb01.html

Carnell, Brian, "China's One Child Policy," May 17, 2000. As of July 23, 2010:
http://web.archive.org/web/20010516010636/http://www.overpopulation.com/faq/Population_Control/one_child.html

Chatterji, Somnath, Paul Kowal, Colin Mathers, Nirmala Naidoo, Emese Verdes, James P. Smith, and Richard Suzman, "The Health of Aging Populations in China and India," *Health Affairs*, Vol. 27, No. 4, July/August 2008, pp. 1052–1063.

"China Hit by Brain Drain, Report Says," *China Daily*, 2007. As of October 7, 2007:
http://www.chinadaily.com.cn/china/2007-06/01/content_884824.htm

"China Shipbuilding Industry Report, 2006–2007." As of November 2009:
http://www.okokok.com.cn/Abroad/Class121/Class102/200705/117803.html

"China Solicits Public Opinions on 12-Year Plan for Education," Xinhua News Agency, January 8, 2009.

"China South Industries Group Corporation Announces Earnings Results for the Year 2008," *Business Week*, January 9, 2009.

"China's Corporation in Technology Intensive Industries!" thread on Chinadaily Bulletin Board System, October 27, 2009. As of March 4, 2011:
http://bbs.chinadaily.com.cn/viewthread.php?action=printable&tid=651134

"China's Defense Budget to Grow 14.9% in 2009," *China Daily News*, Beijing, March 4, 2009. As of November 2009:
http://www.chinadaily.com.cn/china/2009-03/04/content_7535244.htm

"China's Defense Budget to Grow 7.5 Percent in 2010: Spokesman," Xinhua News Agency, March 4, 2010.

"China's Defense Spending to Increase 7.5 Pct in 2010: Draft Budget," Xinhua News Agency, March 5, 2010.

"China's 2010 Military Spending 1.5 Times Larger Than Defense Budget," Kyodo, July 8, 2010. As of January 2011:
http://www.japantoday.com/category/world/view/chinas-2010-military-spending-15-times-larger-than-defense-budget

Chow, Gregory C., and Kui-Wai Li, "China's Economic Growth: 1952–2010," *Economic Development and Cultural Change*, Vol. 51, No. 1, 2002, pp. 247–256.

Cook, Ian G., and Trevor J. B. Dummer, "Changing Health in China: Re-Evaluating the Epidemiological Transition Model," *Health Policy*, Vol. 67, No. 3, March 2004, pp. 329–343.

Cook, Sarah, and Marty Chen, "China and India: Comparative Overview," paper presented at Research Design Workshop on Informality, Poverty, and Growth: Labor Markets in China and India, Harvard University, April 2–3, 2007.

Cooper, Richard N., *Global Public Goods: A Role for China and India United Nations*, Industrial Development Organization, 2005a.

———, "Whither China?" *Japan Center for Economic Research Bulletin*, September 2005b.

Crane, Keith, Roger Cliff, Evan S, Medeiros, James C. Mulvenon, and William H. Overholt, *Modernizing China's Military: Opportunities and Constraints*, Santa Monica, Calif.: RAND Corporation, MG-260-1-AF, 2005. As of July 21, 2010:
http://www.rand.org/pubs/monographs/MG260-1.html

Cui Shiyang, *China: Opportunities, Challenges and Market Entry Strategies*, U.S. Commercial Service and U.S. Consulate General Chengdu, 2007.

Dahlman, Carl J., and Jean-Eric Aubert, *China and the Knowledge Economy Seizing the 21st Century*, Washington, D.C.: World Bank, 2001.

Das Gupta, Monica, Jiang Zhenghua, Li Bohua, Xie Zhenming, Woojin Ching, and Bae Hwa-Ok, "Why Is Son Preference So Persistent in East and South Asia? A Cross-Country Study of China, India, and the Republic of Korea," *Journal of Development Studies*, Vol. 40, No. 2, December 2003, pp. 153–187.

DaVanzo, Julie, Harun Dogo, and Clifford Grammich, "Demographic Trends, Policy Influences, and Economic Effects in China and India Through 2025," Santa Monica, Calif.: RAND Corporation, forthcoming.

David, Henry P., "Eastern Europe: Pronatalist Policies and Private Behaviour," *Population Bulletin* Vol. 36, No. 6, 1982, pp. 1–49.

"Defence Expenditure Increase in Proportion with Growth: India," Defence Talk website, February 17, 2010. As of December 2010:
http://www.defencetalk.com/defence-budget-increase-india-24241/

Dernis, Hélène, Dominic Guellec, and Bruno van Pottelsberghe de la Potterie, "Using Patent Counts for Cross-Country Comparisons of Technology Output," OECD *STI Review*, Vol. 27, 2001, pp. 129–146.

Dernis, Hélène, and Mosahid Khan, "Triadic Patent Families Methodology," STI Working Paper 2004/2, Paris: Organisation for Economic Co-operation and Development, 2004. As of July 21, 2010:
http://www.olis.oecd.org/olis/2004doc.nsf/LinkTo/NT00000EA2/$FILE/JT00160184.pdf

Desai, Prashant, Richard Fairgrieve, Dippanker S. Haldar, A. P. Parigi, and R. Subramanian, *Tapping into the Indian Consumer Market*, McKinsey Global Institute, 2007.

Dollar, David, Mary Hallward-Driemeier, and Taye Mengistae, *Investment Climate and International Integration*, Washington, D.C.: World Bank, Working Paper Series 3323, 2004.

Donohue, Patrick, "Creating Markets for the Base of the Pyramid," BRINQ, October 28, 2009. As of July 21, 2010:
http://www.brinq.com/workshop/archives/2009/10/28/creating-markets-base-pyramid

Dyson, Tim, "The Preliminary Demography of the 2001 Census of India," *Population and Development Review*, Vol. 27, No. 2, June 2001, pp. 341–356.

————, "India's Population—The Future," in Tim Dyson, Robert Cassen, and Leela Visaria, eds., *Twenty-First Century India: Population, Economy, Human Development, and the Environment*, New York: Oxford University Press, 2004, pp. 74–107.

Dyson, Tim, and Pravin Visaria, "Migration and Urbanization: Retrospect and Prospects," in Tim Dyson, Robert Cassen, and Leela Visaria, eds., *Twenty-First Century India: Population, Economy, Human Development, and the Environment*, New York: Oxford University Press, 2004, pp. 108–129.

Eberstadt, Nicholas, "Power and Population in Asia," *Policy Review*, No. 123, February–March 2004. As of July 11, 2010:
http://www.hoover.org/publications/policyreview/3439671.html

————, "Aging in Low-Income Countries: Looking to 2025," in Augusto Lopez-Claros, Michael E. Porter, and Klaus Schwab, eds., *Global Competitiveness Report 2005–2006*, Houndmills, England: Plagrave Macmillan, September 2005, pp. 163–178.

Edlund, Lena, Hongbin Li, Junjian Yi, and Junsen Zhang, *More Men, More Crime: Evidence from China's One-Child Policy*, Institute for the Study of Labor (IZA), IZA Discussion Papers 3214, 2007.

FactSet Mergerstat, LLC, Mergerstat M&A Database, 2009.

Falk, Martin, "What Drives Business R&D Intensity Across OECD Countries?" paper presented at the DRUID Tenth Anniversary Summer Conference 2005, Copenhagen, Denmark, June 27–29, 2005.

Fan, Emma Xioaqin, and Jesus Felipe, "The Diverging Patterns of Profitability, Investment, and Growth of China and India, 1980–2003," Australian National University Working Paper 22/2005, 2005.

Finkelstein, David M., and Kristen Gunnes, eds., *Civil-Military Relations in Today's China: Swimming in a New Sea*, Armonk, N.Y.: M. E. Sharpe, Inc., October 2006.

Fogel, Robert, "$123,000,000,000,000*," *Foreign Policy*, January/February 2010. As of July 210, 2010:
http://www.foreignpolicy.com/articles/2010/01/04/123000000000000

Forecast International, "China (PRC): Section 1—Data," in *International Military Markets—Asia, Australia & Pacific Rim*, 2009a.

————, "India: Section 1—Data," in *International Military Markets—Asia, Australia & Pacific Rim*, 2009b.

Fortin, Nicole M., "Gender Role Attitudes and Women's Labor Market Participation: Opting-Out, AIDS, and the Persistent Appeal of Housewifery," unpublished manuscript, June 2009. As of July 21, 2009:
http://www.econ.ubc.ca/nfortin/Fortin_Gender.pdf

Fromlet, Hubert, "India Versus China: Who Will Be the Winner in the Long Run?" *Economic and Financial Review*, Vol. 12, No. 3, 2005, pp. 111–143.

Gadhok, Taranjot Kaur, "Risks in Delhi: Environmental Concerns," *GIS Development*, no date. As of July 21, 2010:
http://www.gisdevelopment.net/application/natural_hazards/overview/nho0019.htm

Gallagher, Michael, Abrar Hasan, Mary Canning, Howard Newby, Lichia Saner-Yiu, and Ian Whitman, "OECD Reviews of Tertiary Education: China," Paris: Organisation for Economic Co-operation and Development, 2009. As of July 20, 2010:
http://www.oecd.org/dataoecd/42/23/42286617.pdf

Garver, John, *Protracted Contest: Sino-Indian Rivalry in the Twentieth Century*, Seattle and London: University of Washington Press, 2001.

Gereffi, Gary, Vivek Wadhwa, Ben Rissing, and Ryan Ong, "Getting the Numbers Right: International Engineering Education in the United States, China, and India," *Journal of Engineering Education*, January 2008.

Ghosh, Amiya Kumar, *How to Review the Defence Budget*, Islamabad, Pakistan: Pakistan Institute of Legislative Development and Transparency (PILDAT), June 2009.

Gillingham, Robert, and Daniel Kanda, *Pension Reform in India*, International Monetary Fund, WP/01/125, September 2001. As of July 23, 2010:
http://www.imf.org/external/pubs/ft/wp/2001/wp01125.pdf

Ginarte, J., and W. Park, "Determinants of Patent Rights: A Cross-National Study," *Research Policy*, Vol. 26, 1997, pp. 283–301.

Global Economic Forum, website, no date. As of August 17, 2010:
http://www.weforum.org/en/index.htm

GlobalSecurity.org, "China's Actual Defense Budget," March 2010. As of July 21, 2010:
http://www.globalsecurity.org/military/world/china/budget.htm

Goldman, Charles A., Krishna B. Kumar, and Ying Liu, *Education and the Asian Surge: A Comparison of the Education Systems in India and China*, Santa Monica, Calif.: RAND Corporation, OP-218-CAPP, 2008. As of July 23, 2010:
http://www.rand.org/pubs/occasional_papers/OP218.html

Golley, Jane, and Rod Tyers, "China's Growth to 2030: Demographic Change and the Labor Supply Constraint," Australian National University, 2006.

Gordon, David, *The Next Wave of HIV/AIDS: Nigeria, Ethiopia, Russia, India, and China*, Washington, D.C.: National Intelligence Council, September 2002.

Government of India, Union Budget, 1999–2000 through 2009–2010, undated(a). As of October 2010:
http://indiabudget.nic.in/

———, Union Budget 2009–2010, undated(b). As of September 9, 2010:
http://indiabudget.nic.in/ub2009-10/download_index.htm

———, *Thirteenth Finance Commission 2010–2015*, Volume II: *Annexes*, December 2009, pp. 379–380.

Government of India, Department of Science and Technology, National Science and Technology Management Information System, undated. As of March 4, 2011:
http://www.nstmis-dst.org/

Govindarajan, Vijay, and Chris Trimble, "Is Reverse Innovation Like Disruptive Innovation?" *Harvard Business Review* "HBR Now" blog, September 30, 2009. As of July 23, 2010:
http://blogs.harvardbusiness.org/hbr/hbr-now/2009/09/is-reverse-innovation-like-dis.html

Grant, Jonathan, Stijn Hoorens, Suja Sividasan, Mirjam van het Loo, Julie DaVanzo, Lauren Hale, Shawna Gibson, and William Butz, *Low Fertility and Population Ageing: Causes, Consequences and Policy Options*, Santa Monica, Calif.: RAND Corporation, MG-206-EC, 2004. As of July 23, 2010:
http://www.rand.org/pubs/monographs/MG206.html

Greenhalgh, Susan, and Edwin A. Winckler, *Governing China's Population: From Leninist to Neoliberal Biopolitics*, Stanford, Calif.: Stanford University Press, 2005.

Grimmett, Richard F., *Conventional Arms Transfers to Developing Nations, 2002–2009*, Washington, D.C.: Congressional Research Service, RL41403, September 10, 2010.

Gupta, Prashant, Rajat Gupta, and Thomas Netzer, *Building India: Accelerating Infrastructure Projects*, Mumbai, India: McKinsey and Company, August 2009. As of July 23, 2010: http://www.mckinsey.com/locations/india/mckinseyonindia/pdf/Building_India_Executive_Summary_Media_120809.pdf

Gupta, S. P., *India: Vision 2020*, New Delhi: Planning Commission, Government of India, December 2002. As of July 26, 2010: http://planningcommission.nic.in/reports/genrep/pl_vsn2020.pdf

Hart, Stuart, and Clayton M. Christensen, "The Great Leap: Driving Innovation from the Base of the Pyramid," *MIT Sloan Management Review*, Vol. 44, No. 1, Fall 2002. As of July 23, 2010: http://sloanreview.mit.edu/x/4415

Havely, Joe, "Environment Pays Price of Progress," CNN.com, May 3, 2005. As of July 23, 2010: http://www.cnn.com/2005/WORLD/asiapcf/04/27/eyeonchina.environment/index.html

Hawksworth, John, and Gordon Cookson, *The World in 2050: Beyond the BRICs: A Broader Look at Emerging Market Growth Prospects*, PricewaterhouseCoopers, 2008.

He, Jiang, Dongfeng Gu, Xigui Wu, Kristi Reynolds, Xiufang Duan, Chonghua Yao, Jialiant Want, Chung-Shiuan Chen, Jing Chen, Rachel P. Wildman, Michael J. Klag, and Paul K. Whelton, "Major Causes of Death Among Men and Women in China," *New England Journal of Medicine*, Vol. 53, No. 11, September 15, 2005, pp. 1124–1134.

He, Wei, "The Declination of the Growth of the National Defense Budget—An Interpretation of China's 2010 National Defense Budget," *Renmin Ribao* Online, March 6, 2010.

Herd, Richard, and Sean Dougherty, "Growth Prospects in China and India Compared," *European Journal of Comparative Economics*, Vol. 4, No. 1, 2007, pp. 65–89.

Heritage Foundation, 2010 Index of Economic Freedom website, 2010. As of August 23, 2010: http://www.heritage.org/index/

Hofman, Bert, and Louis Kuijs, *Rebalancing China's Growth*, Peterson Institute for International Economics, 2007.

Holz, Carsten A., *China's Economic Growth 1978–2025: What We Know Today About China's Economic Growth Tomorrow*, Hong Kong University of Science and Technology, 2005.

Horton, Susan, "The Nutritional and Epidemiological Transitions," IFPRI 2020 Vision conference panelist comments, Bonn, September 4–6, 2001. As of July 23, 2010: http://www.ifpri.org/2020conference/PDF/summary_horton.pdf

Hu, Albert, "Technology Parks and Regional Economic Growth in China," *Research Policy*, Vol. 36, No. 1, February 2007, pp. 76–87.

Huang, Jikun, Linxiu Zhang, Qiang Li, and Huanguang Qiu, *National and Regional Economic Development Scenarios for China's Food Economy Projections in the Early 21st Century*, Chinese Academy of Sciences, December 2003.

Hudson, Valerie M., and Andrea M. den Boer, *Bare Branches: The Security Implications of Asia's Surplus Male Population*, Cambridge, Mass.: MIT Press, 2004.

Hugo, Graeme, "Declining Fertility and Policy Intervention in Europe: Some Lessons for Australia?" *Journal of Population Research*, Vol. 17, No. 2, November 2000.

IISS—*See* International Institute for Strategic Studies.

India and the World: Scenarios to 2025, World Economic Forum, 2005.

India, Planning Commission, "Eleventh Plan–Chapter 8: Science and Technology," Planning Commission, 2007. Available at: http://planningcommission.nic.in/plans/planrel/fiveyr/11th/11_v1/11v1_ch8.pdf.

"India's Arms Imports to Touch $30 bn by 2012: Assocham," August 16, 2010. As of December 2010 http://www.thaindian.com/newsportal/business/indias-arms-imports-to-touch-30-bn-by-2012-assocham_100233279.html

"India's Pollution Crisis," *Economist*, July 17, 2008.

Information Office of the State Council of the People's Republic of China, *China's National Defense in 2008,* Beijing, January 2009.

International Institute for Strategic Studies, "China's Military Expenditures," *The Military Balance 1995/96,* London: Routledge, 1996, pp. 270–275.

———, "Calculating China's Defence Expenditure," in *The Military Balance 2006*, London: Routledge, 2006, pp. 249–253.

———, "East Asia and Australasia," in *The Military Balance 2009,* London: Routledge, 2009, pp. 375–376.

———, "East Asia and Australia," in *The Military Balance 2010*, London: Routledge, 2010a, pp. 377–440.

———, "Reforming India's Defence Industries," in *The Military Balance 2010,* London: Routledge, 2010b, p. 473–478

International Monetary Fund, *World Economic Outlook Database*, Washington, D.C., April 2010a.

———, *World Economic Outlook Database,* October 2010b. As of November 2010: http://www.imf.org/external/pubs/ft/weo/2010/02/weodata/index.aspx

Jane's, "China's Defence Budget—Is the PLA Cooking the Books?" *Jane's International Defence Review*, February 1995.

———, "Defence Budget, China," *Jane's Sentinel Security Assessment: China and Northeast Asia*, September 29, 2008.

———, "China Defence Budget," *Jane's Defence Budgets: China*, March 11, 2009a.

———, "Jane's Sentinel Security Assessment: South Asia; Defence Budget," June 15, 2009b.

Jha, Prabhat, Rajesh Kumar, Priya Vasa, Neeraj Dhingra, Deva Thirachelvam, and Rahim Moineddin, "Low Male-to-Female Ratio of Children Born in India: National Survey of 1.1 Million Households," *The Lancet*, Vol. 367, No. 9506, January 21, 2006, pp. 211–218.

Jiao, Wu, "China Refutes Military Spending Report," *China Daily*, July 10, 2010. As of December 2010:
http://www.chinadaily.com.cn/world/2010-07/10/content_10089867.htm

Joint United Nations Programme on HIV/AIDS, *Report on the Global AIDS Epidemic*, 2006a. As of July 23, 2010:
http://www.unaids.org/en/KnowledgeCentre/HIVData/GlobalReport/2006/default.asp

———, *AIDS Epidemic Update*, 2006b. As of July 23, 2010:
http://data.unaids.org/pub/EpiReport/2006/2006_EpiUpdate_en.pdf

Jones, Charles I., "R&D-Based Models of Economic Growth," *Journal of Political Economy*, Vol. 103, 1995, pp. 759–784.

———, "Growth and Ideas," in Phillipe Aghion and Steven Durlauf, eds., *Handbook of Economic Growth*, Volume 1A, Amsterdam: Elsevier, 2005.

Joseph, Josy, "Farewell to Foreign Arms?" *The Times of India*, August 1, 2010. As of December 2010: http://timesofindia.indiatimes.com/home/sunday-toi/special-report/Farewell-to-foreign-arms/articleshow/6242138.cms

Joshi, Rohina, Magnolia Cardona, Srinivas Iyengar, A. Sukumar, C Ravi Raju, K. Rama Raju, Krishnam Raju, K. Srinath Reddy, Alan Lopez, and Bruce Neal, "Chronic Diseases Now a Leading Cause of Death in Rural India—Mortality Data from the Andhra Pradesh Rural Health Initiative," *International Journal of Epidemiology*, Vol. 35, No. 6, 2006, pp. 1522–1529.

Keim, Geoffrey N., and Beth Anne Wilson, "India's Future: It's About Jobs," Board of Governors of the Federal Reserve System, International Finance Discussion Papers No. 913, November 2007.

Kim, M. Julie, and Rita Nangia, *Infrastructure Development in India and China—A Comparative Analysis*, Pacific Basin Research Center, August 2008. As of July 23, 2010: http://www.pbrc.soka.edu/Resources/Documents/KimNangia.pdf

Kuznetsov, Yevgeny, "International Migration of Talent and Home Country Development: Towards Virtuous Cycle," Knowledge for Development Program, Washington, D.C.: World Bank Institute, 2006a. As of July 23, 2010: http://www.trabajo.gov.ar/seminarios/2006/files/140606redes/Yevgeny%20Kuznetsov_DiasporasJune14Baires1.ppt

———, *Diaspora Networks and the International Migration of Skills: How Countries Can Draw on Their Talent Abroad*, K4D Program, Washington, D.C.: World Bank, 2006b.

Lallemand, Thierry, and François Rycx, "Are Young and Old WorkersS Harmful for Firm Productivity?" Universite Libre de Bruxelles, Working Paper CEB 09-002.RS, 2009.

Lane, Trevor, "In India, Son Preference Declines with Ideal Family Size, but Remains Strong," *International Family Planning Perspectives*, Vol. 30, No. 2, June 2004, pp. 100–101.

Lanjouw, Jean O., and Mark Schankerman, "The Quality of Ideas: Measuring Innovation with Multiple Indicators," Cambridge, Mass.: National Bureau of Economic Research, Working Paper No. 7345, 1999.

Laurent, Clint, *India—Is It the Next China?* Asian Demographics Ltd., 2006.

Lewis, William W., *The Power of Productivity*, Chicago: The University of Chicago Press, 2004.

Li, Changzu, "Jiedu dusheng zunibing [Understanding the only-child soldiers]," 2001.

Linn, Johannes F., *Regional Cooperation and Integration in Central Asia*, Washington, D.C.: Centennial Group, 2006.

Lueth, Erik, "Capital Flows and Demographics—An Asian Perspective," International Monetary Fund, WP/08/8, 2008.

Mason, Andrew, and Ronald Lee, "Reform and Support Systems for the Elderly in Developing Countries: Capturing the Second Demographic Dividend," *Genus*, Vol. 57, No. 2, 2006, pp. 11–35.

McKinsey Global Institute, *The Emerging Global Labor Market*, October 2005.

———, *From 'Made in China' to 'Sold in China': The Rise of the Chinese Urban Consumer*, November 2006.

Ministry of Human Resource Development, Government of India, *Annual Report 2007–2008*, 2008. As of July 23, 2010:
http://education.nic.in/AR/AR2007-08.pdf

Ministry of Science and Technology, Department of Science and Technology, Government of India, homepage, no date. As of August 30, 2010:
http://www.dst.gov.in/

Mohanty, Deba R., "National Defence Budget 2009–2010," Observer Research Foundation, March 7, 2009. As of July 23, 2010:
http://www.orfonline.org/cms/sites/orfonline/modules/analysis/AnalysisDetail.html?cmaid=15970&mmacmaid=15971

———, "Is India Spending Enough on Defence?" *Political and Defence Weekly*, Vol. 9, No. 26, March 30, 2010.

Moss, Trefor, "China Gives No Explanation for Unexpected Slowdown in Defence Spending," *Jane's Defence Weekly*, March 5, 2010.

Murphy, Kevin M., Andrei Shleifer, and Robert W. Vishny, "The Allocation of Talent: Implications for Growth," *Quarterly Journal of Economics*, Vol. 106, No. 2. May 1991, pp. 503–530.

National AIDS Control Organisation, "HIV Sentinel Surveillance and HIV Estimation in India 2007: A Technical Brief," 2007. As of July 23, 2010:
http://www.nacoonline.org/Quick_Links/HIV_Data/

National Bureau of Statistics of China, *China Statistical Yearbook 2005*, Beijing: China Statistics Press, 2005. As of July 23, 2010:
http://www.stats.gov.cn/tjsj/ndsj/2005/indexeh.htm

———, *China Statistical Yearbook 2009*, Beijing: China Statistics Press, 2008.

National Intelligence Council, *Mapping the Global Future*, December 2004. As of July 23, 2010:
http://www.foia.cia.gov/2020/2020.pdf

———, *Global Trends 2025: A Transformed World*, NIC 2008-003, 2008.

National Science Board, *Science and Engineering Indicators 2006*, Arlington, Va.: National Science Foundation, 2006. As of July 26, 2010:
http://www.nsf.gov/statistics/seind06/

———, *Science and Engineering Indicators 2008*, Arlington, Va.: National Science Foundation, 2008. As of July 26, 2010:
http://www.nsf.gov/statistics/seind08/

National Science Foundation, "Brazil, China, India, Russia, and Taiwan, Lead S&E Article Output of Non-OECD Countries," Infobrief, NSF 07-328, September 2007a.

———, *Asia's Rising Science and Technology Strength: Comparative Indicators for Asia, the European Union, and the United States*, Arlington, Va., Special Report, NSF 07-319, August 2007b.

———, "Science and Engineering Degrees: 1966–2006," Division of Science Resources Statistics, Detailed Statistical Tables NSF 08-321, Arlington, Va., 2008. As of July 23, 2010:
http://www.nsf.gov/statistics/nsf08321/

NSF—*See* National Science Foundation.

Nuclear Threat Initiative, "Second Artillery Corps (SAC)," no date. As of July 20, 2010:
http://www.nti.org/db/china/sac.htm

OECD—*See* Organisation for Economic Co-operation and Development.

Organisation for Economic Co-operation and Development, OECD.StatExtracts, online database, undated. As of August 30, 2010:
http://stats.oecd.org/Index.aspx

———, *Manual on the Measurement of Human Resources Devoted to S&T—Canberra Manual*, OECD, Paris, 1995.

———, *Frascati Manual: Proposed Standard Practice for Surveys on Research and Development*, OECD, Paris, 2002. As of August 30, 2010:
http://www.oecd.org/sti/frascatimanual

———, "Patent Database" based on EPO worldwide Statistical Patent Database (PATSTAT), April 2007a.

———, *OECD Science, Technology and Industry Scoreboard 2007 Innovation and Performance in the Global Economy*, Paris, 2007b.

———, *OECD Employment Outlook 2007*, Paris, 2007c.

———, *OECD Science, Technology and Industry: Outlook 2008*, Paris, 2008a.

———, *Main Science and Technology Indicators Vol. 1/2*, Paris, October 2008b.

———, *Main Science and Technology Indicators Vol. 2/2*, Paris, October 2008c.

O'Neill, Jim, Dominic Wilson, Roopa Purushothaman, and Anna Stupnytska, "How Solid Are the BRICs?" Goldman Sachs Global Economics Paper No. 134, 2005.

Paltsev, Sergey, and John Reilly, "China and India in Energy Markets and Its Implication for Global Greenhouse Gas Emissions, Massachusetts Institute of Technology," 2007a.

———, "Energy Scenarios of East Asia: 2005–2025," Massachusetts Institute of Technology, 2007b.

Pandey, Abhishek, Alok Aggarwal, Richard Devane, and Yevgeny Kuznetsov, *India's Transformation to Knowledge-Based Economy: Evolving Role of the Indian Diaspora*, Evalueserve, July 21, 2004.

Poddar, Tushar, and Eva Yi, "India's Rising Growth Potential," Goldman Sachs, 2007.

Poncet, Sandra, "The Long Term Growth Prospects of the World Economy: Horizon 2050," CEPII Working Paper No. 2006-16, 2006.

Poston, Dudley L., Jr., "Son Preference and Fertility in China," *Journal of Biosocial Science*, Vol. 34, No. 3, July 2002, pp. 333–347.

Poston, Dudley L., Jr., and Peter A. Morrison, "China: Bachelor Bomb," *International Herald-Tribune*, September 14, 2005.

PricewaterhouseCoopers, "How Big Will the Chinese Consumer Market Get by 2025?" July 2007.

Purushothaman, Roopa, "India: Realizing BRICs Potential," Goldman Sachs Global Economics Paper No. 109, 2004.

Quan, Heng, "Income Inequality in China and India: Structural Comparisons," 2006. As of July 23, 2010:
http://www.asianscholarship.org/asf/ejourn/articles/Quan%20Heng2.pdf

Rai, L.P., Naresh Kumar, and S. Madan, "Structural Changes in S&T Research in India," *Scientometrics*, Vol. 50, 2001, pp. 313–321.

Registrar General and Census Commissioner of India, Census of India, 2001. As of July 23, 2010:
http://www.censusindia.net

ResearchInChina, *Company Study of China State Shipbuilding Corporation (CSSC), 2009*, 2009. As of March 4 2011:
http://www.researchinchina.com/htmls/Report/2009/5775.html

Rodrik, Dani, and Arvind Subramanian, "Why India Can Grow at 7 Percent a Year or More: Projections and Reflections, International Monetary Fund," WP/04/118, 2004.

Romer, Paul M., "Endogenous Technological Change," *The Journal of Political Economy*, Vol. 98, No. 5, Part 2: The Problem of Development: A Conference of the Institute for the Study of Free Enterprise Systems, October 1990, pp. S71–S102.

Samuelson, Paul, *Newsweek*, September 19, 1966.

Schaaper, Martin, "Measuring China's Innovation System: National Specificities and International Comparisons," OECD Science, Technology and Industry Working Papers, 2009/1, OECD publishing, 2009.

Shiyang, Cui, *China: Opportunities, Challenges and Market Entry Strategies*, Chengdu, China: U.S. Commercial Service and U.S. Consulate General, 2007.

Shuzhuo, Li, and Jin Xiaovi, "Uxorilocal Marriage in Contemporary Rural China," *Chinese Cross Currents*, Vol. 1, No. 4, October 2004, pp. 64–79.

Silberglitt, Richard, Philip S. Antón, David R. Howell, Anny Wong, Natalie Gassman, Brian A. Jackson, Eric Landree, Shari Lawrence Pfleeger, Elaine M. Newton, and Felicia Wu, *The Global Technology Revolution 2020, In-Depth Analyses: Bio/Nano/Materials/Information Trends, Drivers, Barriers, and Social Implications*, Santa Monica, Calif.: RAND Corporation, TR-303-NIC, 2006. As of July 23, 2010:
http://www.rand.org/pubs/technical_reports/TR303.html

Simanis, Erik, "At the Base of the Pyramid," *Wall Street Journal*, October 26, 2009.

Singh, Manmohan, speech at the Council on Foreign Relations, Washington, D.C., November 24, 2009.

State Council Information Office, *China's National Defense in 2008*, Appendix V, "Defense Expenditure of the PRC," undated.

Stockholm International Peace Research Institute, Military Expenditures Database, undated(a). As of November 2010:
http://milexdata.sipri.org/

———, SIPRI Arms Transfers Database, undated(b). As of December 2010:
http://www.sipri.org/databases/armstransfers

Surry, Eamon, *An Estimate of the Value of Chinese Arms Production*, United Kingdom: Stockholm International Peace Research Institute, July 2007. As of July 23, 2010:
http://www.sipri.org/research/armaments/production/publications/unpubl_aprod/chinese_aprod

Tyers, Rod, Jane Golley, and Iain Bain, "Projected Economic Growth in China and India: The Role of Demographic Change," Australian National University, 2006.

UNAIDS—*See* Joint United Nations Programme on HIV/AIDS.

UNESCO—*See* United Nations Education, Scientific, and Cultural Organization.

United Nations, "China: Instrument for Standardized International Reporting of Military Expenditures (Actual Outlays, Current Prices)" in *Objective Information on Military Matters, Including Transparency of Military Expenditures: Report of the Secretary-General*, 2007, 2008, and 2009.

United Nations Education, Scientific, and Cultural Organization, UNESCO Institute for Statistics Data Centre, website, no date. As of July 23, 2010:
http://stats.uis.unesco.org/unesco/TableViewer/document.aspx?ReportId=143&IF_Language=eng

———, *Manual for Statistics on Scientific and Technological Activities*, 1984. As of July 23, 2010:
http://www.uis.unesco.org/template/pdf/s&t/STSManualMain.pdf

United Nations Population Division, "World Population Prospects: The 2008 Revision Population Database—Panel 1: Basic Data," March 2009a. As of July 26, 2010:
http://esa.un.org/unpp/index.asp

———, "World Population Prospects: The 2008 Revision Population Database—Panel 2: Detailed Data," March 2009b. As of July 26, 2010:
http://esa.un.org/unpp/index.asp?panel=2

University of Pennsylvania, Wharton, "What's Driving India's Rise as an R&B Hub?" Knowledge@Wharton, 2009. As of July 23, 2010:
http://knowledge.wharton.upenn.edu/special_section.cfm?specialID=40

UNPD—*See* United Nations Population Division.

Uppal, Sharanjit, and Sisira Sarma, "Aging, Health and Labour Market Activity: The Case of India," *World Health and Population*, Vol. 9, No. 4, 2007. As of July 23, 2010:
http://www.longwoods.com/product.php?productid=19516

U.S. Census Bureau, International Data Base, 2010. As of February 7, 2010:
http://www.census.gov/ipc/www/idb/

U.S. Central Intelligence Agency, *World Factbook, 2009*. As of July 23, 2010:
https://www.cia.gov/library/publications/the-world-factbook/

U.S. Department of Defense, *Military Power of the People's Republic of China 2009*, report to Congress, Washington, D.C., 2009. As of July 23, 2010:
http://www.defenselink.mil/pubs/pdfs/China_Military_Power_Report_2009.pdf

U.S. Department of Energy, Energy Information Administration, *International Energy Outlook 2008*, June 2008. As of July 23, 2008:
http://www.eia.doe.gov/oiaf/archive/ieo08/index.html

van Zeebroeck, Nicolas, Bruno van Pottelsberghe de la Potterie B., and Dominique Guellec, "Claiming More: The Increased Voluminosity of Patent Applications and Its Determinants," *Research Policy*, Vol. 38, 2009, pp. 1006–1020.

Virmani, Arvind, *A Tripolar Century: USA, China and India*, Indian Council for Research on International Economic Relations, Working Paper No. 160, 2005.

Visaria, Leela, "Mortality Trends and the Health Transition," in Tim Dyson, Robert Cassen, and Leela Visaria, eds., *Twenty-First Century India: Population, Economy, Human Development, and the Environment*, New Delhi: Oxford University Press, 2004, pp. 32–56.

Vonderheid, Erica, "India's Brain Drain May Reverse Flow," Institute of Electrical and Electronics Engineers, July 1, 2002. As of October 7, 2007:
http://www.ieee.org/portal/site/tionline/menuitem.130a3558587d56e8fb2275875bac26c8/index.jsp?&pName=institute_level1_article&TheCat=2201&article=tionline/legacy/INST2002/jul02/findia.xml&

Wadhwa, Vivek, "Losing Our Lead in Innovative R&D," *Business Week*, June 10, 2008.

———, "A Reverse Brain Drain," *Issues in Science and Technology*, Spring 2009a.

———, "Why China's Chip Industry Won't Catch America's," *Business Week*, September 3, 2009b.

———, "India Is Morphing into a Global R&D Hub, but Can It Ever Take on Silicon Valley?" *Tech Crunch*, November 14, 2009c.

Wadhwa, Vivek, Gary Gereffi, Ben Rissing, and Ryan Ong, "Where the Engineers Are," *Issues in Science and Technology*, Spring 2007.

Wadhwa Vivek, Ben Rissing, Hary Gereffi, John Trumpbor, and Pete Engardio, "The Globalization of Innovation: Pharmaceuticals—Can India and China Cure the Global Pharmaceutical Market?" Kauffman Foundation, Working Paper Series, 2008.

Wadhwa, Vivek, AnnaLee Saxenian, Richard Freeman, Gary Gereffi, and Alex Salkever, *America's Loss Is the World's Gain: America's New Immigrant Entrepreneurs*, Kauffman Foundation, March 2009.

Wang, Feng, and Andrew Mason, "The Demographic Factor in China's Transition," *Chinese Journal of Population Science*, 2006.

Wang, Shaoguang, "Estimating China's Defence Expenditure: Some Evidence from Chinese Sources," *The China Quarterly*, No. 147, September 1996, pp. 889–911.

———, "The Military Expenditure of China, 1989–98," in Stockholm International Peace Research Institute, *SIPRI Yearbook 1999*, New York: Oxford University Press, 1999, pp. 334–349.

Whalley, John, and Weimin Zhou, *Technology Upgrading and China's Growth Strategy to 2020*, IAPER Working Paper No. 21, March 2007.

WHO—*See* World Health Organization.

Wikipedia, "Economy of India Foreign Direct Investment," undated. As of February 8, 2011: http://en.wikipedia.org/wiki/Wiki.Economy_of_India-foreign_direct_investment

———, "Military Budget of the People's Republic of China" 2010. As of July 26, 2010: http://en.wikipedia.org/wiki/Military_budget_of_the_People%27s_Republic_of_China

Wilson, Dominic, and Roopa Purushothaman, "Dreaming with BRICs: The Path to 2050," Goldman Sachs Global Economics Paper No. 99, October 2003. As of July 23, 2010: http://www2.goldmansachs.com/ideas/brics/brics-dream.html

Wilson, Dominic, and Anna Stupnytska, "The N-11: More Than an Acronym," Goldman Sachs Global Economics Paper No. 153, 2007.

Winters, Alan, and Shahid Yusuf, eds., *Dancing with Giants: China, India, and the Global Economy*, Washington, D.C.: World Bank Institute of Policy Studies, 2007.

Woetzel, Jonathan, *Preparing for China's Urban Billion*, McKinsey Global Institute, 2008.

Wolf, Charles, Jr., Brian G. Chow, Scott Harold, and Gregory S. Jones, *China's Foreign Investments: Costs and Risks, Benefits and Opportunities*, Santa Monica, Calif.: RAND Corporation, MG-968-OSD, forthcoming.

Wolf, Charles, Jr., K. C. Yeh, Anil Bamezai, Don Henry, and Michael Kennedy, *Long-Term Economic and Military Trends 1994–2015, The United States and Asia*, Santa Monica, Calif.: RAND Corporation, MR-627-OSD, 1995. As of July 23, 2010:
http://www.rand.org/pubs/monograph_reports/MR627.html

World Bank, World Development Indicators online database, no date. As of July 23, 2010:
http://data.worldbank.org/data-catalog/world-development-indicators

———, *Sustaining India's Service Revolution*, Washington, D.C., 2004.

———, *Cost of Pollution in China: Economic Estimates of Physical Damages*, Washington, D.C., 2007. As of July 23, 2010:
http://siteresources.worldbank.org/INTEAPREGTOPENVIRONMENT/Resources/China_Cost_of_Pollution.pdf

———, "Girls' Education," January 4, 2009. As of July 23, 2010:
http://web.worldbank.org/WBSITE/EXTERNAL/TOPICS/EXTEDUCATION/0,,contentMDK:20298916~menuPK:617572~pagePK:148956~piPK:216618~theSitePK:282386,00.html#why

———, Doing Business Indicators website, 2010a. As of July 23, 2010:
http://www.doingbusiness.org

———, "Knowledge Assessment Methodology (KAM)," May 2010b. As of July 23, 2010:
http://web.worldbank.org/WBSITE/EXTERNAL/WBI/WBIPROGRAMS/KFDLP/EXTUNIKAM/0,,menuPK:1414738~pagePK:64168427~piPK:64168435~theSitePK:1414721,00.html

World Economic Forum, *China and the World: Scenarios to 2025*, 2006. As of July 21, 2010:
http://www3.weforum.org/docs/WEF_Scenario_ChinaWorld2025_Report_2010.pdf

———, "The Global Competitiveness Report 2009–2010," Geneva, 2009. As of July 26, 2010:
http://www.weforum.org/en/initiatives/gcp/Global%20Competitiveness%20Report/index.htm

World Health Organization, WHO Statistical Information System (WHOSIS), undated. As of July 26, 2010:
http://www.who.int/whosis/en/index.html

———, *World Health Report 1999: Making a Difference*, 1999. As of July 26, 2010:
http://www.who.int/whr/1999/en/

———, *Country Cooperation Strategy at a Glance: India*, 2006. As of July 26, 2010:
http://www.who.int/countryfocus/cooperation_strategy/ccsbrief_ind_en.pdf

———, Regional Office for the Western Pacific, "Country Context, China," 2009. As of July 26, 2010:
http://www.wpro.who.int/countries/2008/chn

World Values Survey Association, World Values Survey website, 2009. As of July 26, 2010:
http://www.worldvaluessurvey.org/

Wu Chong, "China to Build 30 New Science and Technology Parks," Science and Development Network (SciDev.Net), April 19, 2006. As of July 26, 2010:
http://www.scidev.net/en/news/china-to-build-30-new-science-and-technology-parks.html

Wu, Yanrui, "Service Sector Growth in China and India: A Comparison," *China: An International Journal*, Vol. 5, No. 1, March 2007, pp. 137–154.

Yin, Lu, "Relativity of Military Transparency," *China Daily*, October 29, 2009. As of January 2011:
http://www.chinadaily.com.cn/opinion/2009-10/29/content_8865633.htm